A SMOKING GUN:
HOW THE TOBACCO INDUSTRY GETS AWAY WITH MURDER

*For Jim,
With best
wishes

Beth Ahh
6/89*

OTHER BOOKS BY ELIZABETH M. WHELAN

The 100% Natural, Purely Organic, Cholesterol-Free, Megavitamin, Low-Carbohydrate Nutrition Hoax (with F.J. Stare, M.D.)	1983
Nutrition During Pregnancy	1982
The Expectant Parents' Survival Guide	1982
The Pregnancy Experience	1978
Eat OK — Feel OK (with F.J. Stare, M.D.)	1978
Preventing Cancer: What You Can Do to Reduce Your Risks by Up to 50%	1978
Boy or Girl?	1977
A Baby? . . . Maybe	1975
Panic in the Pantry: Food Facts, Fads and Fallacies (with F.J. Stare, M.D.)	1975
Making Sense Out of Sex (with S.T. Whelan, M.D.)	1975
Sex and Sensibility	1974

A SMOKING GUN: HOW THE TOBACCO INDUSTRY GETS AWAY WITH MURDER

Elizabeth M. Whelan, Sc.D., M.P.H.
Executive Director
American Council on Science and Health

A People's Health Library Book
Edited by
Stephen Barrett, M.D.

George F. Stickley Co.   210 West Washington Square
Philadelphia, PA 19106

The People's Health Library is a series of easy-to-read books written by experts who explain health and health care concepts for the general public. For a complete list of titles, send a self-addressed, stamped envelope to the George F. Stickley Company, 210 West Washington Square, Philadelphia, PA 19106.

Copyright © 1984 by Elizabeth Whelan
ISBN-0-89313-039-7

Library of Congress Card 84-050685

All Rights reserved. No part of this book may be reproduced or used in any form or by any means—graphic, electronic, or mechanical, including photocopying, recording, taping, or information storage and retrieval systems—without permission from the publisher.

Manufactured in the United States of America; Published by the George F. Stickley Company, 210 W. Washington Square, Philadelphia, PA 19106

Contents

Foreword by Luther L. Terry, M.D.   ix
Preface   xiii

PART I: AN AMERICAN TRAGEDY
1. The paradox of cigarettes in the health-conscious '80s   1
2. Tobacco *or* health: the risks of smoking   9
3. Fourteen ploys that can kill you   15

PART II: A BACKWARD GLANCE
4. In the beginning, there were no cigarettes   28
5. A triumph of American ingenuity   40
6. When Camels became kings   47
7. The cigarette hit parade: 1920-1940   56
8. Luckies go to war   72
9. The evidence mounts   81
10. The saga of the '60s   97
11. Showdown in Marlboro country   109
12. Joe Califano and the politics of smoking in the '70s   120
13. Is there a safe tobacco product?   133

PART III: MANDATES FOR ACTION
14. Reflections on a burning issue   141
15. Smokers should carry their load   146
16. "Sue the bastards!"   154
17. Exploitation of developing countries should be ended   166
18. Cigarette advertising should be stopped   177
19. Nonsmokers should be protected   191
20. The smokescreen must be lifted   201
21. Smoking cessation: an overview   212

APPENDIX A. Tobacco's industrial network   225

APPENDIX B. Helpful organizations   228

APPENDIX C. Recommended reading   230

INDEX   234

About the Author

Dr. Elizabeth M. Whelan is co-founder and Executive Director of the American Council on Science and Health, a New York-based not-for-profit, tax exempt consumer education group directed by a panel of some 100 American physicians and other scientists. Dr. Whelan is a graduate of Connecticut College, received a Master's degree from the Yale School of Medicine, and Master's and Doctoral degrees from the Harvard School of Public Health. She is a frequent contributor to both popular and professional publications and one of the nation's leading proponents of the question "smoking *or* health?" She is married to attorney Stephen T. Whelan; they and their daughter, Christine, live in Manhattan.

About the Editor

Stephen Barrett, M.D., a practicing psychiatrist and lecturer on consumer health, is author/editor of 16 books, including *The Health Robbers* (a comprehensive exposé of quackery), *Vitamins and "Health" Foods— The Great American Hustle,* and the college textbook *Consumer Health—A Guide to Intelligent Decisions.* An expert in medical communications, he is Editor of *Nutrition Forum* newsletter and is a scientific and editorial advisor to the American Council on Science and Health.

Acknowledgments

First and foremost, I would like to thank Dr. Paul Magnus, Medical Associate at the National Office of Heart Research in Australia, for his encouragement and criticism—and for providing access to his extensive research materials on smoking and health. Dr. Magnus is one of a very small number of individuals who have focused in depth on the magnitude of the health chaos caused by cigarettes and are dedicated to speaking out about it.

Thanks are due to Cathy Becker Popescu, Research Associate at the American Council on Science and Health, for the sometimes tedious library research for this work and for her major contribution to the chapters of this book dealing with litigation, nonsmokers' rights, the economic impact of cigarette smoking, and the tragedy of cigarettes in the Third World. Lee Francis and Cheryl Martin helped greatly with their research and typing efforts. And Dr. Stephen Barrett did a superb job of editing the manuscript on his legendary IBM word processor.

I am also indebted to Federal Trade Commissioner Michael Pertshuk; William Rothbard, FTC Attorney Advisor; John M. Pinney, President of John M. Pinney Associates and former Director of the U.S. Office of Smoking and Health; Donald Shopland, Technical Information Officer for the U.S. Office of Smoking and Health; John Banzhaf, Executive Director of Action on Smoking and Health; Elizabeth Fayad, Associate Director of the Coalition for Smoking or Health; Matt Myers, Staff Director of the Coalition for Smoking or Health; Dr. Marvin Kristein, Consultant in Health Economics at the American Health Foundation; Steve Wieckert, Legislative Aide to Congressman Thomas Petri; Dr. Mike Daub, Department of Community Medicine, Edinburgh University; and A. Lee Fritschler, Director of the Advanced Study Program at the Brookings Institution and author of *Smoking and Politics*.

A special acknowledgment is due to Martha Lindsey, who typed the many drafts of this book and became committed to the need to alert Americans to the health disaster called the cigarette.

Finally, I heap thanks and praise on my husband Steve Whelan who read each draft of this book, and to my 6-year-old daughter Christine who patiently awaited my emergence from the "writing room" during the summers of 1982 and 1983. I knew that she too was immersed in the subject matter when, in the course of my research, my great aunt died in her mid-80s and Christine asked, "Did Aunt Mary smoke?" When I told her Aunt Mary did not smoke, her puzzled response was "Well, then, why did she die?"

Foreword

When I was appointed U.S. Surgeon General in 1961, like some 50 percent of American men, I was a smoker. Unlike most Americans, however, I was aware of the dangers of smoking even before I took it up. One of my professors in medical school had been Dr. Alton Ochsner, one of the first scientists to investigate the relationship between cigarette smoking and lung cancer. Nevertheless, during my residency training, I somehow slipped into the smoking habit.

Despite my knowledge, I continued to smoke for many years without making a serious attempt to quit. Like many other smokers, I was physically and psychologically dependent upon cigarettes and found it easier to suppress my knowledge rather than give them up. As Surgeon General, however, I felt that continuing to smoke would do more than endanger my own health. Since I was a "role model," my smoking would jeopardize the health of millions of other Americans.

It was difficult, but I finally gave up cigarettes around the time that President Kennedy asked me to appoint an expert advisory committee to resolve the "smoking and health question." I tried switching to a pipe, but soon found that I was smoking it in much the same manner as cigarettes—so I stopped altogether.

The advisory committee met for the first time in November, 1962. They met in secret, but there was no doubt about the conclusion they would reach because the scientific evidence against cigarettes was by that time overwhelming. Their verdict was announced on January 11, 1964. On that day, the committee members and I sat on a platform in the State Department auditorium and confronted reporters with the grim news that smoking was a primary cause of lung cancer, was overwhelmingly associated with emphysema, chronic bronchitis and cardiovascular disease, and was a major factor in premature death.

That evening there was camera coverage of our report on every television outlet, and the next morning, there was front page coverage in every newspaper in the country. During the next several weeks, millions of people gave up smoking. Sales dropped. And we were jubilant! As sensible physicians, public health officers, educators and scientists, we imagined for a moment that we had "conquered" cigarette smoking.

We were wrong. Not altogether wrong, but mostly wrong. Within a few months, millions of those who quit began to smoke again. During the 20 years which have passed, the prevalence of smoking has edged downward, and we can be thankful for that. But some 56 million Americans still smoke over 600 billion cigarettes every year. And more than 350,000 Americans die prematurely each year as a result.

We were unduly hopeful in 1964 for several reasons.

- We vastly underestimated the dependency factor involved in cigarette smoking. Our 1964 report called smoking a "habit" and said that smokng is compulsive with *some* smokers. Today, we know that cigarette smoking is compulsive for *many* if not *most* smokers. Smoking is unquestionably a form of drug addiction just like heroin use. Yet, while possession and selling of heroin is a felony, we allow cigarettes to be passed out on the street and advertised in newspapers, magazines and billboards.

- We also underestimated the capacity which humans have for ignoring and denying unpleasant information. Although surveys indicate that 90 percent of Americans know that cigarette smoking is hazardous, many are only vaguely aware of the magnitude and specifics of the risks involved. Every day, smokers are inundated with cigarette ads which suggest that lots of healthy, young, good-looking people smoke. Rarely do they come across information on smoking's dangers. Thus, it is relatively easy for most smokers to ignore the truth.

- We overlooked another thing in our innocence in 1964: the economic clout of the tobacco industry. Our earlier battle to eliminate polio had been extremely successful, and we expected similar success with cigarette smoking. But there had been no mighty industry nurturing and promoting the polio virus! Tobacco is grown in 22 states and is our sixth largest cash crop. Moreover, a complex network extends the chain of economic dependence to a broad spectrum of other industries. The tobacco industry carries a great deal of weight in business and legislative circles.

- Finally, we overlooked our greatest ally in combatting the cigarette industry: the large numbers of Americans who didn't smoke. In 1964, there were 115 million adults—and 68 million (59 percent) of them did not smoke cigarettes. The percentage of nonsmokers has grown so that today, 67 percent of American adults—108 million people—are nonsmokers.

It never occurred to the authors of the 1964 report to consider the discomfort, annoyance and the actual physical harm which smoking causes nonsmokers. The 1972 Surgeon General's report was the first to review the health effects of "involuntary smoking," but the subject has still not been studied sufficiently.

In recent years, there has been increasing awareness of the problems caused by "second-hand smoke." We are finally beginning to get responsible and sensible no-smoking policies for public areas and places of employment. Aside from protecting nonsmokers, perhaps the greatest value of these policies is their message that smoking should no longer be socially acceptable. As smokers are increasingly asked to refrain or are segregated into smoking sections, young people will be less likely to think smoking is "cool." Smokers, too, may be forced to re-evaluate the wisdom of continuing their habit.

Looking back over the past two decades, there is much good news. Although 55 million Americans continue to smoke, 34 million others who used to smoke have given it up. Per capita cigarette consumption is now at its lowest level in over 30 years. We can also cheer at some recent victories in establishing reasonable no-smoking policies.

We must now devise ways to continue and accelerate this progress. To do this, we must first know what we are up against. This is where *A Smoking Gun* comes in. Examining the history of the cigarette from its "birth" in 1884 to the present time, Dr. Whelan explores the political, economic and social forces which have made the tobacco industry so powerful—and smoking our leading cause of premature death.

I hope this book makes you angry as it describes how the tobacco industry has been hurting *you*—whether or not you are a smoker. I hope it will also inspire you to take action against our number one health enemy: the cigarette.

Luther L. Terry, M.D.

Dr. Terry, who is Emeritus Professor of Research Medicine at the University of Pennsylvania School of Medicine, was Surgeon General of the U.S. Public Health Service from 1961 to 1965.

Preface

The year 1984 marks two important anniversaries in the history of cigarettes. On April 30, 1884, the modern cigarette was born in the North Carolina factory of W. Duke and Sons. On that date the new Bonsack cigarette-rolling machine passed its final test, operating successfully for a full workday. This marked the end of the inefficient handrolling era, and allowed cigarette manufacturers' output—and the number of smokers—to increase substantially. The dramatic but ultimately tragic success story of the cigarette was about to be written.

On January 11, 1964 the cigarette's Golden Age unofficially came to an end with the release of the first U.S. Surgeon General's Report on Smoking and Health. Scientific evidence of the hazards of cigarette smoking had been accumulating for decades, but never before had the facts been summed up so concisely or so effectively. In light of today's knowledge, the report seems mild, but in 1964 it shocked the country and established the link between smoking and disease in the public consciousness.

Today, 20 years after the first Surgeon General's report, we know that smoking is even more dangerous than was believed in 1964. The impact which the cigarette has had on America's health has been far greater—and more deleterious—than anyone could have imagined. Yet we as a country seem to have become resigned to the presence of the cigarette, lacking any commitment or direction in dealing with a consumer product that causes the premature death of more than 350,000 Americans every year. That's nearly 1,000 deaths per day!

The grim statistics related to cigarette smoking are sometimes so overwhelming they blur the reality. To put the 350,000 mortality toll in perspective, it is useful to recall that when two jumbo jets collided on the runway of Los Rodeos airport at Tenerife in the Canary Islands on March 28, 1977, 580 people lost their lives. Media coverage of the event was nothing short of spectacular. *Newsweek* dedicated six pages to the incident. The thought of so many human lives being tragically extinguished stirred people around the world and prompted journalists to talk of the "collision course" and "deaths in the afternoon" in articles carrying illustrations of airplane hangars filled to capacity with coffins. *If every single day two filled-to-capacity jumbo jets crashed—killing all on* board—the death toll (about 212,000) would not approach that accounted for each year by cigarette smoking. Yet, ironically, we have no media-moaning and no outrage over the cigarette carnage, and the "smoking gun" in question is the subject of nearly $1.5 billion in advertising which promotes it as an essential part of the sophisticated, smart, macho, ambitious, successful, fun-loving, outdoorsy, robust, profile.

How can a product as deleterious as the cigarette exist in this health-conscious society, a country apparently committed to reducing risks of premature death? How did the cigarette secure its grip on our society and why, in light of the frightening health consequences of smoking, is that grip not loosening? Is there a solution to the problem of the cigarette in America?

A Smoking Gun will closely examine the phenomenon of the cigarette, its spectacular marketing and advertising success, the floundering attempts of medical professionals, the media, advertisers, tobacco men, Congress, and the general public to deal with newly discovered health hazards of smoking. *A Smoking Gun* is the attempt of one public health professional to oppose the juggernaut of the cigarette "political-industrial complex." It is a chronicle of the politics of the cigarette, the frantic activity of those dependent on tobacco for profit—and those who represent these people in Congress—to save the cigarette no matter what it does to human health. It is, necessarily, an examination of the darker, more perverse aspect of human nature, one which allows, and fosters, the marketing of a lethal product—for the sake of one's own economic avarice.

But most of all, *A Smoking Gun* is a quest to answer a very basic and sobering question: *How can we, in the 1980s, come to grips with the devastating health and economic consequences of cigarettes and still keep our actions compatible with a free enterprise system that respects individual freedom?*

Clearly, our first step must be to face up to the enormity of the problems that cigarettes now pose. We must cast aside that odd sense of resignation and recognize cigarettes for what they are: a unique problem that merits a unique solution. Let us, as a country, make a policy decision one way or another on cigarettes. Given our experience with Prohibition in the 1920s—an attempt of government to deny access to a commodity that people want—the outlawing of cigarettes seems like an unrealistic option. But we can remove our heads from the sand and face the ultimate question: Do we choose to continue dispersing the enormous costs of smoking throughout the entire population? In other words, will it be the country's policy decision to keep cigarettes basically unregulated and continue to force all Americans to pay the bill for smoking-related disability and death? Are we willing to accept the cigarette as such an essential part of American economics and lifestyle that we consider the costs in dollars and lives tolerable? If only for historical purposes, we have the moral and ethical duty to confront this dilemma and decide on a course of action. And if our decision is to proceed on our current track, attended only by our well-meaning but not very effective attempts at "education," it may well go down in the history books as a strong indictment of our political and economic way of life.

This book is not a tirade against cigarettes. It is not an attempt to scare the wits out of you by categorizing all the gruesome things that can happen to you if you smoke. It is not evangelical and, hopefully, does not moralize. The book makes no claim to being "neutral," but then there is no more reason to be neutral about cigarettes than there is to be neutral about malaria or drunk driving. The book does offer the framework of an essential two-part plan to deal with the cigarette tragedy, a structure which I feel is miles away from "public health do-gooding" and is fully consistent with my own commitment to freedom of choice, free enterprise and the all-American way:

1. *Smokers and nonsmokers must be given all the facts about the health consequences of smoking.* National surveys indicate that Americans do not understand the risks involved in smoking. Because of the tobacco industry's grip on the U.S. media and other channels of communication, a free flow of bad news information about smoking has been inhibited. Freedom of choice is only truly valid when the choice is fully informed.

2. *The true costs of smoking should be shifted to the tobacco industry and the smokers.* Given that the cigarette is the only product legally available in the United States which is harmful when used as intended, the industry should be held financially responsible for the damage done by their product. This cost, of course, should ultimately be passed on to their customers; and the smoker, not nonsmokers, should bear the enormous costs of cigarette-induced disease and property damage.

If you are a smoker, the book should anger you as you realize that you are being victimized by a multi-billion dollar industry which thrives by marketing a deadly product. You should feel rage over the ploys the tobacco industry uses to distract your attention from the devastation of cigarettes and the effort it makes to keep you hooked on cigarettes. As the victim of this manipulation, your outrage should cause you to fight back, to resist their propaganda and to hold the tobacco companies legally responsible for your current physiological addiction to cigarettes and damage to your health.

If you are a nonsmoker, you should also be outraged by the costs you are assuming because of a smoker's "right" to smoke. You should be indignant about the fact that you pay higher insurance premiums, higher taxes and social security contributions because of the phenomenally high health costs resulting from cigarette-induced disease.

And you might want to give some considerable attention to the question of "nonsmokers' rights" to clean, breathable air and to airplanes, apartment houses and hotels free from the risk of cigarette-induced fires.

As a good citizen, you should be outraged by the outrageous activities of an industry that is literally getting away with murder through its ability to control so many segments of American society, all in the interest of perpetuating a life-threatening enterprise. You will find that the whole topic of the cigarette in America is complex, intriguing—and in some senses—intimidating. The fact that only a handful of books have been published on this topic for the public during this century suggests that it is not a subject that authors eagerly pursue. And for good reasons. Mysterious things happen when you start traveling into cigarette country. It sometimes seems that the tobacco industry has a fleet of guardian angels (or as the Senator quoted below refers to them, "godfathers") making sure that the bad news about cigarettes is kept to a minimum.

For example, consider a perceptive and insightful quote from *Smoke Screen,* a dynamic and enlightening book on the politics of tobacco by Former Senator Maurine B. Neuberger (D-Oregon), a quote with which I personally identify. She wrote:

> Though I ceased believing in witches and goblins and the like when I was a young girl . . . I confess that my study of the tobacco problem has greatly shaken my disbelief. How is one to explain the extraordinary frequency with which some unidentified force has intervened to prevent the public from learning about the hazards of smoking, without concluding that the tobacco industry is protected by a benign fairy godfather?

Senator Neuberger tells the story of a distinguished scientist who was asked to join in a one-hour film for network television. In the two minutes allotted him, he spoke about the dangers of cigarettes, using a British poster depicting a smoldering cigarette whose twisting string of smoke spelled out "Cancer." Was it simply coincidence that the producers later reported this was the only part of the film destroyed in processing?

The Senator also describes the experience of a pharmaceutical magazine editor who wrote a gripping editorial on the role of tobacco in disease. Being a compulsive editor, he checked his manuscript until it was in final version—galley proofs—to make sure it was just as he wanted it. The proofs were satisfactory, but during the intervening hours before the magazine was printed, his editorial was garbled in such a way as to distort his message.

Senator Neuberger would surely not have been surprised to learn that in 1976, after Britain's Thames Television aired a half-hour documentary about six cowboys who developed lung cancer or emphysema, Philip Morris, manufacturer of Marlboro, was successful in getting a court order stopping the circulation of the film.*

---

*The film "Death in the West" was ordered destroyed, but somehow a copy of the tape showed up in May of 1982 and was aired twice on KRON-TV in San Francisco. It has since been aired on several other local networks around the country,

Over the course of the six years I have been intensively studying the problem of tobacco and health, I have come to sense the presence of many invisible backstage hands orchestrating the successful effort to keep the health effects of tobacco off-stage, out of sight of a worried audience. One or two unexplained events, of course, are not convincing. But when they are brought together, the conclusion is unavoidable: the tobacco industry does have a godfather—indeed thousands of them—on full-time alert to squelch any attempt to spotlight the health devastation associated with tobacco.

On Tuesday, March 21, 1978, I was to appear on a live segment of an extremely popular network television program to discuss my newly published book *Preventing Cancer* (W.W. Norton, 1978). A primary conclusion of that book was that cigarette smoking accounted for more than one-third of all cancer deaths in the United States.

The appearance was announced in advance in *TV Guide* and the *New York Times*. I had been a guest a number of times before on this program, and I knew the routine: the night before I awaited the network's call to learn what time the limousine would pick me up (a courtesy offered to all program guests). Instead, I received a curt call from a producer at 7:20 P.M. indicating, without reason, that my segment was being cancelled. The next morning, the program, lacking a live guest, ran a taped filler.

In March 1980, the group which I co-founded and now direct, The American Council on Science and Health (ACSH), held a news conference in San Francisco to release the results of its intensive study and evaluation of the extent of coverage given to smoking and health by the nation's major women's magazines. With the help of a public relations agency, we invited some 300 members of the radio, television and print media. In addition, we mailed out some 200 press releases to media outlets outside San Francisco.

Our study's findings were by anyone's definition newsworthy: While smoking had indeed "come a long way, baby," and was the leading cause of preventable death in women, the women's magazines ignored the topic. What we had uncovered, of course, was the fact that the heavy advertising revenues (over $1.2 billion in 1980, with most of it to newspapers and magazines) derived from tobacco companies had caused magazines to suppress discussion of the hazards of cigarettes. (The exceptions, including some general audience magazines discussed later, are *Reader's Digest, Good Housekeeping,* and the *New Yorker,* none of which accept cigarette ads and all of which point out the dangers of smoking on a regular basis.)

To present our findings and their implications for public health, we brought together a distinguished panel of scientists, including the Dean of the School of Public Health at the University of California, Berkeley.

---

always receiving a favorable audience response. The national networks, apparently afraid of retaliation by an advertising "godfather," have declined to broadcast it.

xviii   *Preface*

While most ACSH press conferences have filled a small room to capacity, only six press representatives showed up, four from radio, two from television. There were no newspaper or magazine reporters there, no newspaper picked up the press release we mailed out—and the stories I had given in advance to East Coast media contacts never found their way into print. The tobacco godfather was on the job.

He was there too in the many instances when my manuscripts were "edited" by magazine editors. On one occasion I prepared an article on cancer and the environment for *Harper's Bazaar*. I began with a section entitled "lung cancer" which emphasized the guilt of tobacco in producing this disease. When the article appeared in print, the first section was "Breast Cancer"—and the section on lung cancer was buried in the back of the magazine with the classified ads. Another time an article of mine on the safety of natural and processed foods was accepted by *Signature*. At some point I referred to cigarette smoking as a major cause of cancer. When the manuscript appeared in print, the line read "heavy cigarette smoking," an obvious effort to tone down my statement and reassure pack-a-day smokers who would never think of themselves as "heavy smokers."

The tobacco godfather is often subtle and behind the scenes. But he can get pretty heavy-handed, too. In September 1980, I was invited by the Pineapple Association of America to address a group of food editors in Honolulu on the topic of Food and Nutrition in the 1980s. I agreed to do this, and wrote my talk. Two days before my departure for the Islands, I received a call from the San Francisco-based trade association which was to be my host; and the caller told me in plain language: "While you are our guest in Honolulu we ask that you kindly refrain from talking of the health hazards of tobacco." I was instantly perplexed, given first that I was to talk about nutrition and had no intention of talking about tobacco and second, I couldn't figure out the relationship of pineapple and tobacco. *Then* he spelled it out: "One of our association's major members is Del Monte Pineapple owned by R.J. Reynolds and they would not appreciate your anti-smoking sentiments."

On July 4, 1980, the American Council on Science and Health declared "Independence from Cigarettes" Day in New York City, complete with speakers and a rally in midtown. A few months before the event we asked Mayor Edward Koch to issue a proclamation acknowledging this event. He agreed—and then a few weeks later decided it would not be appropriate.

What this godfather discussion boils down to is the stark reality that an enormous number of people in the United States today are "dependent" on cigarettes in some way or another. Major food companies are owned by tobacco conglomerates, and are reluctant to be outspoken about the role of tobacco in disease. Newspapers and magazines are so intimately in-

Preface xix

volved with tobacco interests, that as any staff writer or freelance contributor knows—and if they don't know, they are told up-front—cigarettes and health are not fair game for discussion.* While television and radio no longer carry cigarette ads, most of the commercially sponsored television shows do carry copious ads from companies owned by tobacco interests; so the influence is there. Thus, a television or radio show that carried ads for 7-UP, which is owned by Philip Morris, would not be inclined to focus in detail on the relationship of cigarettes and disease. Municipal governments are dependent on cigarette advertising revenues in buses, subways and other public places. Bank officers are mum about tobacco's health effects because they handle money from some very wealthy tobacco clients.

Even some of the people whom you would think would want to disseminate information about cigarettes, don't. For example, pharmacists might be ideal health educators on the hazards of tobacco if they accepted that role. But they don't because they sell cigarettes, and thus have their own economic interest in them. Large chemical companies might seem likely candidates to lead the fight against tobacco if for no other reason than to clear themselves and their chemicals as the cause of cancer. But they don't because many of them make products (like fertilizers, curing agents) which they in turn sell to the tobacco companies.

Individuals, because of their own self interest—sometimes even though they are not personally in favor of the use of cigarettes—have bound together in a silent allegiance to perpetuate the cigarette and downplay or ignore its impact on health. The commitment to silence is omnipresent, and if you don't believe it, try this: attempt to begin a discussion of the relationship of cigarettes and disease at a cocktail party or other social gathering. You will find that there is immediately tension in the room and probably some instant attempts to change the subject. You would surely have more success in starting a discussion if you introduced the topic of leprosy. Americans today do not want to talk about cigarettes and may pass it off with "everyone knows that, let's discuss something else." But the time has indeed come that we must discuss it. As the old campaign slogan goes: if not now, when?

I wrote a good deal of this book on the Southern Shore of New Jersey, gazing at the thundering Atlantic Ocean. Sometimes the ocean was calm,

---

*Two other personal examples may be of interest. *Harper's Bazaar* in the late 1970s asked me to do a piece called "Protect Your Man From Cancer." I did it and by necessity discussed the role of tobacco in lung, bladder, pancreatic and oral cavity cancers. The piece was returned—with a check for full payment—saying that they had too many cigarette ads to allow such a piece. When a colleague of mine and I were regularly contributing a column to *Family Weekly* we were told not only to avoid mention of cigarettes, but also to avoid the topic of cancer, lest we inadvertently say something offensive to advertisers.

but there were occasions when it reminded me in no uncertain terms of its power. There were times when, having finished my daily regimen of swimming, I'd turn to swim to shore and found that I could not. The power of the ocean was such that struggling, at least for the moment, seemed useless. There have been times in preparing this manuscript that I have felt a similar feeling of helplessness and hopelessness, aware of my own personal insignificance in a sea of tobacco interests. In the face of a seemingly hopeless dilemma, it is tempting simply to give in, which in this case, would mean joining the conspiracy of silence myself. But recalling the words of Marshal Foch, I persevered: "There are no hopeless situations; there are only men who have grown hopeless about them."

Elizabeth M. Whelan, Sc.D., M.P.H.

1

The Paradox of Cigarettes in the Health-Conscious '80s

She (at breakfast): "Well, now I'm going to find out who quit smoking recently."
He: "How do you do that?"
She: "I just read the obituary page."

When it comes to attitudes about health and the environment, Americans seem to be guided by a double standard. On one hand, near hysteria prevails in discussions about dioxin in Missouri, EDB residues in grain products, alleged radioactive contamination at Three Mile Island, urea formaldehyde foam in home insulation, nitrite in bacon, and various other food additives. While in some of these cases *suggestive* evidence indicates a *potential* threat and just reason for vigilance, no solid evidence exists that any of the above situations has actually caused death or serious disease in human beings.

On the other hand, although hundreds of thousands of Americans die prematurely each year from diseases associated with cigarette smoking, we hear only an occasional protest about this undeniable environmental hazard.

How can a product as dangerous as the cigarette be tolerated with relative calm in our health-conscious and demanding society? Why are there are not frequent television documentaries, investigative reports, outraged citizen groups and concerned legislators crusading against what is clearly the most dramatic and far-reaching public health threat of all time? The answer lies in a malevolent web of five factors:

1. Cigarettes became an established part of American life before the extent of their harmfulness was apparent.

2. The cigarette habit can be extremely difficult to give up because cigarettes are physically addicting and psychologically habit-forming.

3. People don't like to be reminded that they are killing themselves.

4. The tobacco industry spends enormous amounts of money trying to reassure smokers that smoking is safe, pleasurable and socially acceptable.
5. The tobacco industry still has tremendous political clout.

Backfire

The cigarette represents a cruel backfire of human innovation. *If cigarettes were being considered for introduction today, there is no way they could meet the safety criteria of either the Food and Drug Administration or the Consumer Product Safety Commission.* Burned tobacco contains a significant number of cancer-causing agents, and the inhalation of tobacco smoke produces immediate adverse effects upon the body.

But cigarettes are *not* just being introduced; they have been around for a hundred years. For 60 of these years, honest, hard-working, enterprising Americans grew tobacco, manufactured cigarettes, and promoted and distributed the product in a clever and well-intentioned manner. The period from 1890 to 1950 marked the golden age of the cigarette as it came to symbolize the all-American man and eventually the all-American woman. Controversy and rumblings of danger existed, but by the time the bad news was clear, cigarettes had become socially desirable, and large segments of our economic system—including the United States government—had become dependent on cigarettes as a source of revenue.

Mental and physical dependence

Scientific studies indicate that the nicotine in tobacco is just as addictive as heroin! Beyond the physical addiction, there is the social/psychological dependence on the smoking habit. For many smokers, the behavior modification needed to stop smoking presents more of a barrier than the physical addiction. (For example, many people who give up smoking "don't know what to do" with their hands.)

Dr. William Polin, Director of the National Institute on Drug Abuse, terms cigarette smoking "the most widespread drug dependence in our country." The average smoker consumes 30 cigarettes per day. Each inhaled puff delivers a dose of nicotine to the brain, resulting in 50,000 to 70,000 doses per person each year. No other form of drug use occurs with such frequency or regularity.

National surveys indicate that 90 percent of smokers would like to quit, but 85 percent who have tried have failed. Cigarette smokers are physiological prisoners, their bodies craving nicotine in order to remain comfortable. When smokers say they have tried to stop but can't, they really mean it. And of those who do stop smoking, some 70 percent resume the habit within three months—about the same recidivism rate as with heroin.

Psychological blackout

Clearly, the addictive and habit-forming qualities of cigarette smoking are keys to the tobacco industry's survival. But its survival is also classic testimony to the existence of psychological mechanisms that protect people from unpleasant facts with which they cannot cope. When new information clashes with longstanding beliefs, a mental reaction called "cognitive dissonance" occurs. This was vividly evident during the 1950s when epidemiological studies began to find that smokers had extraordinary rates of lung cancer, heart disease, emphysema and other ailments. Most Americans simply refused to incorporate this new information into their consciousness. Throughout that decade there was enormous resistance by journalists, legislators and even physicians to believing the new findings. During the early '50s, almost 70 percent of American men were smokers. Most of them had thought the habit was at worst only slightly harmful; they simply could not cope with a different assessment.

Today's surveys show that 90 percent of Americans know cigarette smoking is hazardous to health (although most of them "know" it only in a rhetorical sense and are unaware of the specific dangers). But coping mechanisms to reduce cognitive dissonance still play a role. Smokers must cope with the reality that they smoke and with the available evidence that it is extremely harmful. Since two such beliefs cannot coexist comfortably, the evidence of harm is suppressed and in some cases sublimated. (The latter is illustrated by the smoker who declares his "health consciousness" by joining an exercise club or shopping at a "health food" store.)

In plain English, smokers prefer not to talk or think about the dangers of their habit. Public service advertisements about the hazards of smoking are few and far between, so most smokers today find it relatively easy to avoid mental dissonance. Cigarette smoking is an automatic form of behavior, with no incentive for smokers to re-evaluate their decision to smoke from week to week.

It would seem logical that nonsmokers, unencumbered by the physiological addiction to cigarettes, would be an ideal source of encouragement for smoking cessation. But again, it generally does not work that way because of some basic human psychological mechanisms. First, disease data on smoking are so horrifying that it may be difficult for a wife, for example, to allow herself to even imagine her husband experiencing an excruciating death from lung cancer. A type of "second-hand" denial may set in. Second, discussion of the subject might cause marital discord. Third, many nonsmokers sense the smoker's depression, guilt and unhappiness—and fear that they will make a bad situation worse by bringing up a topic that is a real "downer." Others may be seriously concerned about the health of their smoking friends or loved ones but simply baffled about how to bring up the topic tactfully and effectively without irritating the person they are trying to help. How odd that cigarette smoking is more socially acceptable than discussions of its health hazards!

Economic and political clout

Cigarettes are big business in the United States. With 600 billion cigarettes smoked annually at a cost of close to $20 billion, the economic stakes are very high. Not only is tobacco grown in 22 American states, making it the sixth largest cash crop, but there is a vast and complex tobacco supply network which extends the chain of economic dependence on tobacco to include a full spectrum of industries, including manufacturers of farm supplies and equipment, transportation, advertising—and in turn to those who depend on these suppliers. A further ripple effect extends from tobacco manufacturers to Madison Avenue ad agencies and finally to U.S. newspapers, magazines and other outlets which collect over one billion dollars annually from cigarette advertisers. And any list of "cigarette dependents" must include federal, state and local governments which receive more than $6 billion in cigarette and excise taxes.

There are five major ways that tobacco interests flex their political muscle when they perceive any threat to their beloved products.

1. *They rely on the corporate clout of their family companies.* By buying soft drink, insurance, Chinese food, and alcoholic beverage companies, they have extended the reach of their corporate teeth. Under other circumstances, Del Monte, which operates canneries and specialty plants, beverage operations, seafood and frozen food plants around the world would have no interest in the sales of cigarettes. But as part of the R.J. Reynolds family, Del Monte has a definite interest in the success of their parent company. Similarly, it's all-in-the-family for Miller Beer and Seven-Up, with their "father" Philip Morris; and for Saks Fifth Avenue, Gimbels Department Stores, and Brown and Williamson.

2. *They wield power over their suppliers*—the thousands of American businesses that manufacture the agricultural chemicals, paper, cardboard and other materials they need to make, market and distribute cigarettes. Tobacco executives know that businesses need clients. And the tobacco empire is a very valuable client, one that many businesses would do nothing to displease. Thus major chemical companies which produce agricultural chemicals are part of the tobacco "family" too, because if they become outspoken about the dangers of cigarettes, they might lose these affluent customers. And the roots go deeper. The sobering reality is that not only are the suppliers of cigarette manufacturers themselves beholden to the tobacco interests—but this dependency may well extend to those who deal with the companies *owned* by the tobacco conglomerates—such as those which supply glass and containers for Seven-Up, cans for Miller beer, sugar for Hawaiian Punch, flavoring agents for

Patio Mexican foods and Chung King Oriental food. These suppliers deal indirectly but still significantly with the tobacco empire.

3. *Cigarette companies have political allies.* By teaming up with the manufacturers of other products that might be the subject of bad press or government regulation, the cigarette manufacturers are constantly seeking potential allies who will stand by them in the name of Big Business and free enterprise. Tobacco companies can appeal to a wide variety of corporate entities, suggesting that if government regulation is successful in the tobacco business, it will next move on to their industries. The message here is "let's stick together or we'll all go down together."

An analysis of the affiliation of the directors of the top tobacco companies demonstrates how successful cigarette manufacturers have been in making corporate officers of other industries members of the tobacco team. For example, the R.J. Reynolds board includes Herschel H. Cudd, retired Senior Vice President at Standard Oil, Ronald Grierson from the General Electric Company, John D. Macomber from Celanese Corporation and John W. Hanley of Monsanto Chemical Company.

Morton Mintz, a *Washington Post* staff writer, noted in April 1982 some curious conflicts of interest in the interlocking of corporate directorships. George V. Comfort, Robert E.R. Huntley, William Donaldson and Margaret B. Young are all paid directors of life insurance companies. They also receive at least $20,000 a year as board members of Philip Morris. Believe it or not, two other outside directors of Philip Morris are trustees of medical institutions whose doctors embrace the conventional medical wisdom of the cause-effect relationship of smoking and disease. H. Robert Marschalk, a retired vice chairman of Richardson-Vicks, Inc., is on the board of overseers of Dartmouth Medical School, and John S. Reed, a Senior Executive Vice President of Citicorp in New York is a trustee of Memorial Sloan-Kettering Cancer Center (he can bring to Sloan-Kettering "some of the realism of the cigarette world"). Mintz reported that Reed sees "no anomaly" in this situation. Both he and Marschalk are nonsmokers but told Mintz that health considerations are not involved in that decision.

4. *Cigarette manufacturers demonstrate their economic clout through the use of some of the most elaborate and extravagant advertising and promotional budgets in American history.* Although they publicly deny that cigarettes are devastating to health, there is no possibility that the decision-makers at the big five tobacco companies are unaware of the risk associated with their product. Thus they have made a conscious decision that their own economic well-being is far more important than the health of their customers. Advertising is their primary mechanism for neutralizing the medical fears among smokers and keeping the "pleasures" of smoking in the public's mind.

Cigarette apologists have for years defended their advertising practices by claiming that they advertise only to foster competition among various brands, not to mislead smokers about the effects of their habit, or to lure new smokers to the ranks. But an analysis of the ads makes it obvious that they are promoting not cigarettes, but *social acceptance* of cigarettes. Right now, smokers are understandably very nervous, unsure of the legitimacy of their smoking behavior. Cigarette advertising attempts to reassure by suggesting that smokers are sophisticated, sociable, fun-loving, attractive and healthy. Advertising also minimizes coverage of smoking's dangers by magazines and newspapers which don't wish to risk losing the enormous revenue involved.

5. *Tobacco companies have been highly successful in manipulating Senators and Congressmen* who depend on them for support or who, through some wheeling and dealing and good old-fashioned Congressional horse-trading, can raise support for other issues by compromising themselves on the issue of cigarettes and health.

Cigarette politics in the '80s

During the 1960s, tobacco proponents demonstrated nearly complete control over any tobacco legislation that was introduced. In the '80s, the political stance of the Congressional members of the tobacco substructure has been considerably less formal, simply because the industry has felt considerably less threatened. During the 97th Congress (1980 to 1982), only three major types of cigarette legislation were considered. One, sponsored successfully by Senator Jesse Helms, Washington's Number One Guardian of the Health of the Cigarette Industry, was supposed to make the tobacco subsidy system self-sustaining. Instead, it was basically a public relations gesture which enabled the industry to trade one benefit for another (see Chapter 15). The second aimed to strengthen the warning label on packages and in advertising.

The 97th Congress did raise the excise tax on cigarettes, but although the health issue was mentioned in passing, the primary intent of this bill was to raise revenues for general spending purposes. The main reason that few bills antagonistic to tobacco interests are seriously considered is probably that nearly every relevant subcommittee has high-ranking members from tobacco-producing states who make sure that such legislation is rarely introduced.

The tobacco issue is obviously geographical rather than partisan. In addition to Senator Helms (R-NC), the tobacco substructure on the Hill is dominated by Senators Walter D. Huddleston (D-KY) and John P. East (R-NC), and Representative Thomas J. Bliley, Jr. (R-VA) and Harold Rogers (R-KY).

Tobacco's defenders strongly oppose strengthening the warning labels on cigarette packages and in advertising, primarily because this could prove to be a step in the direction of a total ban on cigarette advertising. Rep. Rogers said a House bill to strengthen warning labels was "overreaching in constitutional aspects [which would be a] brainwashing operation to sway free people on what they should or should not do." Senator Huddleston complained that the bill was "an unwarranted and untested intrusion." Rep. Bliley, a member of the Waxman subcommittee considering the bill, deserves a Broken Record Award for asking, "Shouldn't we get the evidence first?" During the Fall 1982 hearings, he stalled for time until the subcommittee was forced to adjourn.

Tobacco's Congressional substructure wields its power not only by sticking together to defend their constituents' primary cash crop—but by cashing in on I.O.U.s. A substantially greater percentage of black Americans than white Americans are hooked on cigarettes. But the Black Caucus in the House appears to have traded silence on cigarettes for Southern support on civil rights issues.

Vote-swapping has involved other odd bedfellows. Congressman Carl D. Perkins (D-KY), Chairman of the Education and Labor Committee, has been supporting the interests of labor unions and could cash in on his credits with representatives from non-tobacco states who have large labor union constituencies. Congressman Charles Rose (D-NC) was not only a strong supporter of the Congressional vote to bail out New York City during its economic woes, but also supported the government financial rescue of Chrysler, so he had some votes he could count on when he was in trouble.

The politics of tobacco can be complex and intriguing. Congressmen James J. Florio (D-NJ) and Toby Moffitt (D-CT) had both been outspoken about the dangers of pollution in the environment and the need for stricter standards of quality for air, water and workplace settings. Both were members of the House Committee which considered strengthening the cigarette warning bill, but neither actively supported the bill. By repeatedly not showing up for committee meetings and refusing to give their proxy to anyone, they blocked the bill's passage out of the committee. Florio's district includes Camden, New Jersey, whose largest industry is Campbell Soup. Florio actively opposed salt labeling on processed foods, a stand that would make Campbell Soup happy. Speculation in Washington is that Florio did not want to appear inconsistent on the subject of labeling by supporting the bill dealing with cigarettes, but not the one dealing with salt. Moffitt's constituency included many New York advertising executives.

An abrupt reversal of the Reagan Administration on proposals to strengthen the cigarette warning label and increase anti-smoking education provides yet another example of tobacco's influence. The Administration had at first voiced support for Representative Waxman's warning bill. But later the Assistant Secretary of Health and Human Services *retracted* that support, to the surprise of Surgeon General C. Everett Koop who favored the legislation but was told to "back off of the issue" by the Administration. It is not unlikely that President Reagan's close friend, Senator Jesse Helms persuaded him to change his mind on this matter during one of his frequent visits with Mr. Reagan.

Why there is hope

We might as well forget trying to dissuade that united front of representatives dedicated to protecting the economy of their states—no matter what the cost to human health. But that does leave politicians from 46 other states. While there are some lines of dependency on tobacco in these states, there is also tremendous potential for lining up support to confront the tobacco lobbies. Most Senators and Congressmen are rational human beings who are aware that smoking is harmful. What they need is constituent support so that they can stand up for what they believe.

The unique circumstances of cigarettes' survival necessitate a complex and multifaceted approach to deal with this issue. Legislation is sorely needed to prohibit cigarette advertising, promote public education to discourage smoking, inhibit the promotion and sales overseas of U.S.-grown tobacco, transfer the costs of smoking to the smoker, and place the tobacco industry under some type of federal regulation. Tobacco interests are strong and organized. But they *can* be overwhelmed if voters send a clear message that the time has come to face up to and solve the critical health problem of cigarettes.

2

Tobacco or Health: The Risks of Smoking

> *"Worldwide, even the most conservative estimates would place the number of avoidable deaths caused by smoking well over one million annually."*—World Smoking and Health
> An American Cancer Society publication

Virtually everyone today "knows" that cigarettes are hazardous to health. We've heard about the increased incidence of lung, bladder, mouth, esophageal and other cancers, the dramatically increased risks of heart disease and emphysema, and the threat of birth defects in children whose mothers smoke during pregnancy. Every year or so, headlines cite another round of grim statistics from the Surgeon General's office. But the majority of smokers know the risks only rhetorically—*they have no emotional grasp of the enormity of the dangers involved in smoking.*

Cigarette deaths are not just statistical; they involve real people. Edward R. Murrow, Betty Grable, Humphrey Bogart, Walt Disney, Buster Keaton, Nat "King" Cole, and England's King George VI are among those who died of cigarette-induced cancers. Lung cancer does not discriminate, striking down even "beautiful people" in the prime of life. In December 1979, *Fortune* magazine featured Wilhelmina Behmenburg Cooper, the renowned Dutch-born model who came to preside over her own modeling and talent agency. Three months later she died *at the age of 40* of cigarette-induced lung cancer.

A look at the obituary page can be quite revealing. When you read of a man in his mid-40s dying suddenly from a heart attack, or a woman killed at age 59 by emphysema, the odds are that these deaths were cigarette-related.

During the past few years the fathers of a number of my high school friends died of lung cancer. My accountant, diagnosed as having esophageal cancer at age 50, died four short months later, smoking cigarettes with his last breaths. My aunt—the mother of three young children—died from burns received when careless use of cigarettes in bed burned her room beyond recognition. Yes, friends and loved ones suffer, and consciously or not, we know the cause. Perhaps your favorite teacher has dropped dead at age 55 from a sudden heart attack. Or closer to home, a parent—or your spouse—may be wheezing incessantly, considering every staircase a major challenge.

Here is a brief summary of the dangers of cigarette smoking based on government reports.

Overall risks

- Cigarette smoking is the greatest single cause of preventable death in the United States today. Well over 350,000 deaths annually are smoking-related. Four of the five leading causes of death are related to cigarette smoking.
- Regular smokers lose about five minutes of life expectancy for each cigarette they smoke.
- More than one out of every seven deaths in this country is related to smoking. Each year *six times* as many people in this country die from smoking-related causes as die from automobile accidents.
- Overall, a smoker is 70 percent more likely to die at a given age than is a comparable nonsmoker. Heavy smokers are 200 percent more likely to die prematurely than are nonsmokers.
- Smoking two or more packs a day decreases life expectancy more than eight years. One pack a day decreases life expectancy six years.
- The earlier one starts smoking, the more likely one is to die from it.
- Male cigarette smokers report 33 percent more days lost from work and 14 percent more days of bed disability than do men who have never smoked. Female smokers have an absentee rate 45 percent greater than that of nonsmokers and report 17 percent more days of bed disability.

Heart and blood vessel disease

- The nicotine in cigarette smoke can cause increased heart rate and can contribute to high blood pressure. The carbon monoxide in cigarette smoke seems to encourage accumulation of atheromas (the fatty plaques in arterial walls) which contribute to blockage of blood supply to the heart. Carbon monoxide also reduces the amount of oxygen delivered to the cells. At the levels found in cigarette smoke, carbon monoxide is dangerous to people with heart or lung disease.
- Heart disease accounts for nearly half of all deaths in this country. Cigarette smoking accounts for 30 percent of all heart disease deaths. Far more smokers die from heart disease caused by cigarettes than die from cigarette-induced cancers.

Tobacco or Health: The Risks of Smoking 11

- In 1982 some 170,000 heart disease deaths were attributable to cigarette smoking.
- It is estimated that 1 of every 10 Americans alive today will die prematurely of cigarette-related heart disease.
- Smoking is the major avoidable risk factor for heart disease. Smokers are 2 to 3 times more likely to die of heart disease than are nonsmokers.
- Smoking significantly increases the chances of a second heart attack in someone who has already had one.
- Smoking aggravates circulatory diseases of the arms and legs. People with such conditions who continue to smoke increase the risk of gangrene, amputations and even death.
- Women who smoke and use birth control pills have a very high risk of cardiovascular complications. They are 10 times as likely to suffer a heart attack and 20 times as likely to suffer a stroke by cerebral hemorrhage as a woman who does neither.

Lung cancer

- Lung cancer is the leading cause of cancer death in men and is expected to surpass breast cancer in 1984 to become the leading cause of cancer death in women.
- Smoking is the number one cause of lung cancer in this country, accounting for 80 percent of all lung cancer deaths. In 1982, more than 90,000 people died of lung cancer caused by smoking.
- Overall cancer mortality rates among smokers are dose-related as measured by the number of cigarettes smoked per day. Smokers are *10 times* more likely to die from lung cancer than nonsmokers. Very heavy smokers (two packs or more per day) are up to *25 times* more likely to die of lung cancer.
- Few people survive lung cancer. Seventy percent of lung cancer patients die within one year of diagnosis, and 90 percent within five years.

Other cancers

- Cigarette smoking has been established as a significant cause of cancer of the larynx, oral cavity, esophagus and bladder, and is significantly associated with cancer of the pancreas and kidney.
- There is a synergistic relationship between smoking and the use of alcohol which greatly increases the risk of cancer of the larynx, oral cavity and esophagus for those who smoke and drink heavily.
- Some studies have indicated a relationship between cigarette smoking and cancer of the stomach and cervix; but more research is needed to confirm or refute these findings.

Estimated 1984 cancer deaths attributable to smoking are shown in Table 2:1.

Table 2:1. Smoking and cancer death, 1984 estimates

Cancer site	Total deaths	Percent attributable to smoking	Total deaths attributable to smoking
Males			
Lung	85,000	83%	70,550
Mouth, pharynx, larynx, esophagus	12,700	75%	9,525
Bladder	7,300	56%	4,088
Pancreas	12,000	40%	4,800
Total, males	117,000	76%	88,963
Females			
Lung	36,000	43%	15,480
Mouth, pharynx, larynx, esophagus	5,250	43%	2,257
Bladder	3,400	25%	850
Pancreas	11,000	25%	2,750
Total, females	55,650	38%	21,337

Other lung diseases

- Smoking can destroy the cilia, the tiny hairs lining the air passageways which continuously sweep particles away from the lungs. When this happens, susceptibility to infection increases.
- Smoking is the major cause of both emphysema and chronic bronchitis.
- In 1982, there were more than 46,000 smoking-related deaths from emphysema and chronic bronchitis, representing 85 percent of the deaths from these conditions.
- Lung function is measurably impaired in smokers.

Peptic ulcer disease

- Twice as many smokers as nonsmokers die from ulcers of the stomach and duodenum (upper part of the small intestine).
- Smoking appears to retard the healing of peptic ulcers.

Pregnancy and infant health

- Maternal smoking during pregnancy significantly increases the risk of spontaneous abortion (miscarriage).
- Maternal smoking during pregnancy increases the risk of stillbirth or infant death within a month of birth by at least 20 percent for light smokers (less than one pack a day), and by 35 percent for those who smoke more than one pack a day.

- Mothers who smoke are considerably more likely to give birth prematurely.
- Babies born to women who smoke during pregnancy weigh an average of 200 grams (about half a pound) less than babies born to comparable women who do not smoke during pregnancy.

Cigarette-induced fires

- Some 2,000 deaths and 4,000 injuries each year in the United States are due to cigarette-induced fires.

Quantifying the risks

Table 2:2 considers the risks that smokers assume. The most critical element is the fact that *a smoker assumes all these risks at the same time!*

Table 2:2. The risks of smoking

Disease category	Percentage of increased risk to smokers*
Lung cancer	700 to 1500%
Laryngeal cancer	500 to 1300%
Oral cancer	300 to 1500%
Esophageal cancer	400 to 500%
Bladder cancer	100 to 300%
Pancreatic cancer	100%
Kidney cancer	50%
Coronary heart disease	30 to 300%
Emphysema and other chronic disease of airway obstruction (excluding asthma)	1000 to 3000%
Peptic ulcer disease	100%

*A person smoking less than a pack a day would be assuming the risks at the lower end of the spectrum.

Economists currently estimate that more than $11 billion per year is spent on direct medical costs of smoking-related diseases. Most premiums for health and life insurance are higher as a result. Taxpayers must cover the extra cost to the Medicare and Medicaid programs. Workers who smoke take substantially more sick leave, and this burden is passed on to consumers in the form of higher prices. In addition, more than $300 million worth of property is lost each year to fires ignited by cigarettes.

These gloomy facts have a brighter side: long-term smokers who quit decrease their odds of sickness and death. For example, ten years after giving up smoking, one's risk of heart disease approaches that of lifetime nonsmokers. After 15 years, former smokers have lung cancer mortality rates only slightly higher than non-smokers; and the risk of other cancers is similarly reduced.

More than 33 million Americans have succeeded in stopping since the Surgeon General's first report on smoking was issued in 1964. Most current smokers have tried to quit at least once. Standing alone in support of smoking—claiming that the case against cigarettes is unproven and merely "statistical"—is the Tobacco Institute. Its major propaganda ploys are analyzed in the next chapter.

3

Fourteen Ploys That Can Kill You

Tobacco companies' annual reports are filled with statistics, and their marketing strategies are based on them. It's easy to understand why statistics that reflect on financial health are acceptable to the industry while those reflecting on human health are not.

Tobacco companies are ruthless in defending their products and apparently will stop at nothing to remain in the business of peddling their lethal weed. Even a severe critic must credit the industry for its thoroughness. Its representatives seem to monitor every obscure meeting, journal and media appearance to clutch at any straw that might point in the industry's favor. They obviously have medical and epidemiological talent on board because their protests indicate expert understanding of scientific methodology and jargon (combined, unfortunately, with a unique and perverse ability to twist facts).

Headquarters for tobacco propaganda is the Tobacco Institute in Washington, D.C. Though founded during the 1950s to "foster public understanding of the smoking and health controversy," its policy toward adverse facts is like that of the official Soviet news agency, *TASS*!

The Institute publishes *The Tobacco Observer*, a monthly newspaper, as well as a variety of brochures, booklets and articles denying the fact that smoking is a major health hazard. It deals with the press and provides speakers who represent the industry's case to live audiences and on radio and television broadcasts. It also lobbies, supported by the Tobacco Action Network (TAN), a voluntary grass roots organization of individuals involved in the production and sale of tobacco products. TAN is said to be "dedicated to freedom of choice."

The defenses used by cigarette pushers can be divided into two categories: *denial* and *distraction*. The first revolves around the theme, "There is no evidence that cigarettes cause disease." The second is a collection of extraneous arguments which might make sense if the first premise were true. Here are 14 of the industry's leading ploys.

Ploy #1: "The apparent increase in the lung cancer rate may be the result of improved ability to detect that disease."

This so-called "diagnosis" explanation for the increased rates of lung cancer has been around for decades. When the increase in lung cancer was first noted during the 1920s, even physicians wondered the same thing. Pointing to the development of X-ray examinations, bronchoscopy and sputum cytology, today's tobacco apologists ask whether the increase is real. But no scientist outside of the tobacco industry has any such doubts. Techniques for diagnosing lung cancer have improved—but so have techniques for detecting other types of cancer. Only in the respiratory tract have cancers shown a dramatic rise.

Ploy #2: "The cause of the disease is not the smoking but some characteristic of the smoker."

This is another old argument. The idea is to take the heat off cigarettes by claiming the smokers are "different" in ways that might make them prone to higher risks of heart disease and cancer. Industry propaganda states that smokers are "more communicative, more energetic, more prone to drink quantities of black coffee and liquor, and to like spicy or salty food in contrast to blander diets." (This last claim may be related to the fact that most smokers have decreased taste bud sensitivity and need extra flavoring to make food edible.) Smokers are also alleged to have "more marriages, more jobs, more residences, living in what you might call overdrive.... It is possible that [the smoker is] also the kind of person more prone to developing lung cancer."

The "character" explanation simply does not square with the increase in lung cancer noted since 1930 or the fact that smokers who stop have lower death rates. Nor does it fit the fact that the death rates of pipe and cigar smokers are not as high as those of cigarette smokers.

Ploy #3: "It's all statistical; you shouldn't confuse association with cause. Epidemiology deals with statistical relationships and comparisons. It cannot determine cause."

Another tried and quite untrue defense. When it suits their purposes, tobacco companies would have us believe that statistics are just a group of numbers looking for an argument, that "statisticians are people who collect data and draw confusions."

In advancing this defense, the companies are fond of quoting a cautious comment from the 1964 Surgeon General's report: "Statistical methods cannot establish proof of a causal relationship in an association. The causal significance of an association is a matter of judgment that goes beyond any statement of statistical probability." The statement was meant to be applied to limited data, such as single studies or small groups of studies. But the case against cigarettes is documented by over 30,000 references in medical and scientific journals around the world—and is supported by research involving millions of people, thousands of competent researchers and more than four decades of work, all of which have yielded chillingly consistent results. During the past quarter century, no medical or scientific group has ever concluded that the case against cigarettes is unproven.

Dr. Ernst Wynder, a pioneer researcher in the field of tobacco and health, now Director of the American Health Foundation, has proposed three criteria that must be satisfied to establish causation between an environmental factor and human disease. Let's apply them to the data on cigarettes.

1. The greater and more prolonged the exposure to the factor, the greater the risk of the population involved. A substantial number of epidemiological studies have shown that the more cigarettes smoked, the greater the risk of developing cancer. In other words, the relationship is dose-related.

2. The epidemiological pattern should be consistent with the distribution of the factor. Lung cancer mortality rates in the United States as well as in other countries are related to per capita cigarette consumption during the last 40 years. Lung cancer rates in specific population groups also correspond with smoking habits. For example, the fact that Jewish men have fewer cancers than men of other religious groups is consistent with the fact that Jewish men smoke fewer cigarettes. So is the low incidence of lung cancer among Seventh-day Adventists, whose religion forbids smoking.

3. Removal or reduction of the factor for a given population group should be followed by a reduction in the incidence of the disease. Epidemiological studies show clearly that people who stop smoking lower their chances of getting cancer and heart disease.

These postulates can be applied to other cigarette-disease links as well. Don't forget that tobacco companies' annual reports are filled with statistics, and that their marketing strategies are based on them. It doesn't take much imagination to understand why statistics that reflect on financial health are acceptable to the industry while those reflecting on human health are not!

While tobacco pushers are quick to dismiss and belittle the methodology and data that implicate cigarettes, *they have never indicated what type of test protocol or data they would accept as a valid test of cigarettes' harmfulness.* Thus it is quite clear that no data pointing to cigarettes as a hazard would ever be acceptable. In 1982, to explore this issue, I sent the following inquiry to Horace R. Kornegay, Chairman of the Tobacco Institute:

> Dear Mr. Kornegay:
> The American Council on Science and Health (ACSH) has recently completed an extensive review of the literature which links cigarette smoking with various diseases. We have also noted the tobacco industry's commentary to the effect that the topic still remains controversial, that statistics cannot be used as a basis for establishing causal relationships, that the data collected to date are flawed, and that there is a need for more research.
> Certainly, every industry has criteria by which to judge the safety of its products. We would be most appreciative if you or a member of your staff could advise us of the tobacco industry's product safety criteria and answer the following questions:
> A. What type of evidence would the Tobacco Institute accept as reasonable proof that smoking cigarettes significantly increased the risk of disease?
> B. What evidence would convince the tobacco industry that its product was hazardous to human health?
> C. What research methodology or methodologies would the Tobacco Institute suggest to collect the type of data which would generate answers to questions A. and B.?
> D. Finally, what action would the Tobacco Institute and the various tobacco companies take if proof of cigarettes' hazard to health were obtained?
> Thank you very much for your assistance in this matter.

Mr. Kornegay replied as follows:

> Your position regarding tobacco, most recently expressed in your statement to a House subcommittee, is abundantly clear and, all things considered, highly predictable. Consequently, no purpose would be served by engaging in a dialogue with you on scientific standards, methodology, criteria of proof and so forth. These important issues can only be resolved within the framework of our democratic system.

Letters asking the same questions were also sent to the chief executives of the five major American tobacco companies, none of whom replied.

Early in 1984, the R.J. Reynolds Tobacco Company broke with the tobacco industry's unwritten code and launched a multimillion dollar ad campaign claiming that cigarette smoking's role in human disease is still scientifically controversial. Asking "Can We Have An Open Debate About Smoking?" the company's board chairman attempted to soothe the anxieties of smokers by suggesting that the verdict was not yet in—and that most problems relating to cigarettes could be solved if smokers and nonsmokers would simply be more tolerant of each other.

While tobacco interests have never acknowledged that their products are harmful or contain potentially harmful ingredients, they have 1) refused to reveal what chemicals are added during the manufacturing process, 2) introduced filters, and 3) engaged in a continuous battle over whose cigarettes contain the lowest amounts of "tar" and nicotine. If nothing is harmful, why keep it secret, try to filter it out or boast that less of it is present? Alan Blum, M.D., founder of DOC (Doctors Ought to Care), has posed a fascinating question. Can you imagine buying a loaf of bread advertised as having "only 6 mg of poison" or a soup labeled "lowest in cancer-causing agents"?

Ploy #4: "If smoking causes cancer, why doesn't everyone who smokes get it?" (Or the corollary: "How come nonsmokers get cancer too?")

This "all-or-nothing" approach flies in the face of epidemiological logic. No physician or scientist working in the area of smoking and health claims that *everyone* who smokes gets cancer or that tobacco use accounts for *all* cases of lung cancer (although it certainly accounts for at least 80 percent of them). Individual susceptibility to disease obviously varies. Less than two percent of persons infected with polio virus develop paralytic polio. Not everyone exposed to the tuberculosis microbe develops the disease. Yet no one would discount the role of these germs in causing illness. Similarly, not everyone who crosses the street without looking both ways will be hit by a car. Does that mean it is safe to do so? Ploy #4 is especially dangerous because it reinforces the wishful thinking of smokers who tell themselves, "It can't happen to me."

Tobacco men used to claim that cigarettes can't cause cancer, because women, who began smoking in large numbers after World War II, did not immediately demonstrate an increasing rate of lung cancer. However, 40 years later, the rate has risen sharply to the degree that lung cancer will soon pass breast cancer as the leading cause of cancer death in American women.

Ploy #5: "No one factor could cause so many diseases."

The tobacco industry frequently complains, "We are a little weary of being the punching bags for those who look on this as the only health issue in the United States," and that "Never before (or since) has a committee 'discovered' a single cause for so many diseases."

Some of the early researchers themselves were shocked by the extent of adverse effects of cigarettes on the human body. But cigarettes contain thousands of chemicals, including many known to affect the body adversely—and many others whose effects have not even been studied.

Ploy #6: "Blame the other guy."

To divert attention from cigarettes, the tobacco folks love to play up news of other factors that contribute to disease. For example, recent issues of *The Tobacco Observer* have carried headlines like "Stress Called Malady of the Decade," "Less Heart Disease Risk in Stress-Reducing Culture," and "Stress/Lung Cancer Linked."

Articles focus on other possible environmental facts with claims such as, "Study Links Air Pollution, Cancer" and call for more research about the possible cancer-causing effects of automobile exhausts. Since the report on diet and cancer made in 1982 by the National Academy of Sciences, tobacco men have been claiming that to avoid lung cancer, we should eat more foods rich in beta-carotene. While there is evidence that some of the broader aspects of diet may influence the odds of developing certain forms of cancer, this by no means diminishes the fact that cigarettes are the major *known* cause.

Most of all, cigarette-makers like to pin the cancer rap on "chemicals" in the workplace. They provide us with tables listing "chemicals associated with cancer induction in man" (asbestos, arsenic, benzene, chromium, nickel, etc.) and remind us that coal miners, foundry workers, textile workers, leather workers, etc., have higher rates of various forms of cancer. They are fond of quoting former HEW Secretary Joseph Califano's "Estimates Paper" which concluded that 20 to 40 percent of all U.S. cancer is caused by occupational exposure, despite the fact that: 1) it was never published; 2) most of its alleged authors disavowed it; and 3) it has been dismissed as invalid by the scientific community.

Syndicated columnist Jack Anderson supported the tobacco industry's "blame the chemicals" ploy by unjustly accusing the American Cancer Society of "bureaucratic astigmatism." He wrote: "The Cancer Society, critics suspect, doesn't want to endanger its corporate contributions by pointing the finger at industries that pollute the workplace and the environment with carcinogens."

A 1979 in-house memo from British American Tobacco's Brown and Williamson Corporation—leaked to the British press—clearly supports the strategy of using the "emergence of other causes of social concern" as a means of dealing with public anxiety about cigarettes and with proposals for legislative action against tobacco interests. This memo is discussed further in Chapter 18.

Ploy #7: "Not all doctors and scientists..."

The following headlines appeared in just *one* issue of *The Tobacco Observer* (April 1982):
"Health Cost (of smoking) Denied"
"Scientists Doubt Ads Cause Smokers to Start"

"Lung Cancer Remains a Mystery"
"Biological Link Said Lacking in Smoking-Cancer Controversy"
"Heart Disease Label 'Invalid'"
"Smoking Said Not Addictive"

Tobacco men try to belittle the massive amounts of scientific data indicting tobacco as a cause of disease by passing it off only as the work of "some scientists"—meaning just a few. They also use this phrase to bolster their own views—referring to unnamed individuals who may very well not exist.

A few well-credentialed individuals still spout the industry line. Among them are Dr. Theodor D. Sterling, of Simon Fraser University, British Columbia, and Dr. Carl C. Seltzer, senior research associate at the Harvard School of Public Health (and a recipient of tobacco industry research funds). But thousands of similarly prominent scientists have spoken out against cigarettes.

The most interesting point to make in discussing the industry's apologists is that they rarely attempt to present medical or epidemiological data to bolster their contention that cigarettes are not hazardous. The reason is very simple: *there are no such data!* They do, however, sometimes include complicated-looking graphs which look appropriate for a newscast on "Saturday Night Live." An example is shown on next page.

Sometimes the Tobacco Institute misinterprets legitimate research data in an attempt to bolster its position. One of its favorite false claims is that the Multiple Risk Factor Intervention Trial (MRFIT) demonstrates that cigarette smoking does not cause heart disease. This study, reported in the September 24, 1982 *Journal of the American Medical Association,* compared an "intervention" group with a "control" group that did not receive a concentrated educational program aimed at reducing risk factors for heart disease. The incidence of heart disease was similar in the two groups, but not for the reason suggested by the Tobacco Institute. Many individuals in the control group simply stopped smoking on their own. In both groups, former smokers and those who stopped during the study period *did* have fewer heart attacks.

Ploy #8: "Everything these days seems to cause cancer."

"Another day—Another chance something will be found hazardous to my health" is the caption of a typical cartoon in *The Tobacco Observer.* "Living is Inevitably Fatal" is another representative headline.

The message here is that everything these days seems to be harmful, and since we are all going to die anyway, there is no point in worrying about cigarettes. Tobacco-related editorials talk of "cancerphobia," remind us that sunlight is carcinogenic, and ask if there should be a warning in the sky that "Sunlight has been determined by the Surgeon General of the

Cancer Incidence Rates per 100,000 U. S. Population

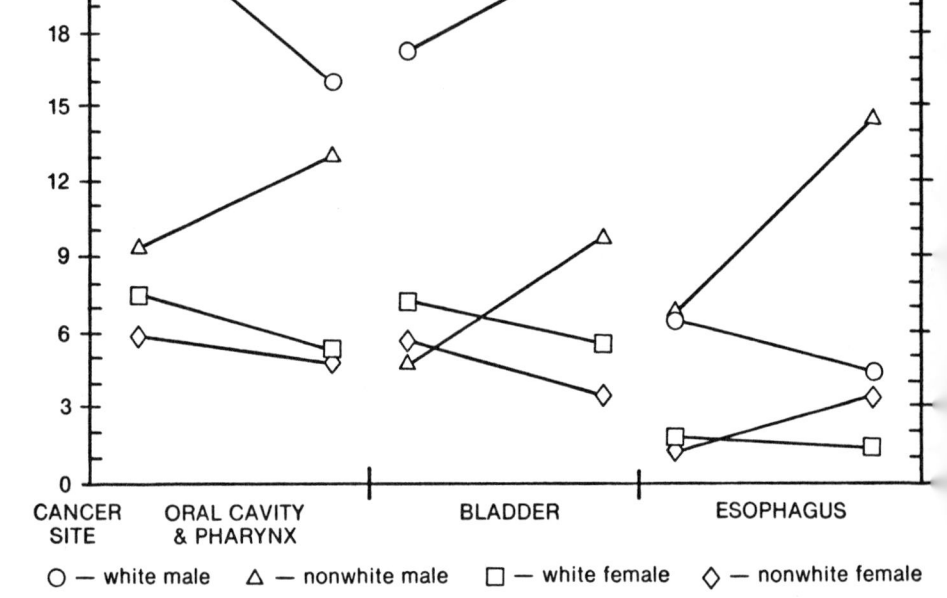

○ — white male   △ — nonwhite male   □ — white female   ◇ — nonwhite female

This chart from the Tobacco Institute's recent review, "Cigarette Smoking and Cancer: A Scientific Perspective," compares the rates of three kinds of cancer as measured by government statistics for the periods 1947-49 and 1969-71. The government links all three of these cancers with cigarette smoking, but the Tobacco Institute asks why the lines don't all go in the same direction. The reason is that these cancers have other causal factors besides cigarette smoking, and that neither these factors nor tobacco use were equally distributed among the four groups during the time periods involved. Thus there is no reason to expect the cancer rates to follow similar patterns, even though they are all related to cigarette smoking.

United States to be Hazardous to your Health." They suggest that cigarettes belong in the company of black pepper (which causes cancer in mice), hair dryers (which contain small, harmless amounts of asbestos), charcoal steaks (the unconfirmed rumor being that the grilling introduces potential carcinogens), peanut butter (which can contain the naturally occurring cancer-causing aflatoxin molds), shampoos, bacon, bubble bath, and a long list of other consumer products. The obvious intent here is to blur cigarettes into the background of other anxieties which the press raises daily about the environment.

Another similar theme relates to the concept of "excess." Tobacco folks acknowledge that anything in excess can be harmful. For example, North Carolina Governor Terry Sanford once said, "I believe that most Americans with any horse sense will recognize that in all things of life, excesses should be avoided—and this applies to our speed in automobiles, our eating and drinking and smoking habits." Another North Carolina Governor, Jim Hunt, said, "There are a number of things that I believe are more harmful—alcohol is one of them and cholesterol and sugar are also harmful in excess." The attempt here is to make cigarettes just part of the crowd of things which are safe if not used "excessively." But "moderate" cigarette smoking is not safe either!

Ploy #9: "Look at all the research we do."

Ever since the industry began speaking about the "cancer scare" in the mid-1950s, "research" has proven to be an excellent public relations tool. Tobacco companies have probably spent close to $100 million for research during the past 30 years. Does that sound like a lot? It is less than one-tenth the industry's annual advertising budget in the United States. Industry publications constantly remind us of the "research"—and seem to seek some prestige for themselves by noting that the money was given to Harvard Medical School, UCLA School of Medicine and other such prestigious institutions. But they rarely reveal when the research points to cigarettes as a health hazard—which it almost always does.

Ploy #10: "The opposition are all weirdos."

The basic tactic here is to suggest that prominent opponents are self-serving "do-gooders" who are "anti-pleasure." For example, former HEW Secretary Joseph Califano was accused of being a "zealot" who simply wanted to convince people that what *he* did (quitting smoking) was the right thing for everybody to do. They publicized cartoons labeling him "Carrie Nation Califano," suggesting falsely that his motivation was simply puritanical.

The Tobacco Institute claims that "antismokers" are "a small minority of nonsmokers—lobbying for laws to restrict or prohibit the use of tobacco, seeking to convert a custom into a crime." Included among these "nonsmoking pests" are Action on Smoking and Health (ASH), Citizens Against Public Smoking (CAPS), The Health Research Group (HRG), Californians for Nonsmokers Rights (CNR), Doctors Ought to Care (DOC), the Group to Alleviate Smoking Pollution (GASP) and—I am proud to say—the American Council on Science and Health (ACSH).

What motivates "antismokers"? The industry posed that question in the *Tobacco Observer* to Dr. Peter Berger, a Boston College sociologist who according to the tobacco folks "has written a number of important books." He replied that anti-smokers are motivated by a desire to "win government recognition and raise more money to stabilize and enhance the different groups they represent," and that they demonstrate a type of "denial of human mortality" especially when they talk about "unnecessary deaths." He also said:

> They call smoking an "epidemic," a "menace." They call their movement a "crusade" and talk of it in militaristic terms such as "enemies," "battles," and "victories." It seems to me that this language is significant in disclosing the fanatical psychology of at least a segment of the anti-smoking movement.

Apparently, by tobacco industry reasoning, no one can be interested in promoting public health without being selfish, self-serving or power-hungry.

Ploy #11: "The concept of 'excess deaths' is unjustified."

The Tobacco men bristle most when they hear the Surgeon General's office claim that cigarette smoking is responsible for over 350,000 "excess deaths."

The primary tactic to defuse the excess death estimates is to claim that they aren't real—that "No one who uses the figures can say accurately where they originated." This is not so. The estimate is based on the well grounded method of comparing "expected" deaths with those observed in epidemiological studies. For example, if you followed comparable groups of 10,000 smokers and 10,000 nonsmokers and found that over a 5-year period there were 123 lung cancer deaths among the smokers, and 20 among the nonsmokers, the number of "excess" deaths from lung cancer would be 103 deaths. Figures for each disease category and age group can be combined to calculate overall and age-specific rates.

Dr. Morton L. Levin, formerly Chief of the Department of Epidemiology of Roswell Park Memorial Institute, began such calculations in 1962, using the mortality rates of males aged 18 and over, sorted into disease categories. He concluded that 83.5 percent of the deaths from lung cancer could be considered excess, as could 76 percent of those from bronchitis and emphysema, 72 percent from cancer of the larynx, 59 percent of those from cancer of the oral cavity, 52 percent of deaths from cancer of the esophagus, 28 percent of coronary heart disease deaths. Overall, 25.7 percent of deaths in male smokers were calculated as excess. In 1966 Dr. Levin concluded that 6.4 percent of deaths in female smokers were excess, and estimated the overall excess mortality due to cigarettes to be about 300,000. He noted:

> This is admittedly a crude estimate and is made with the knowledge that it could be in error by as much as 10 or 15 percent. Nevertheless, it does

provide a rough estimate of the magnitude of the problem presented since it represents the number of cigarette smokers who died in 1966 because cigarette smokers have higher death rates than nonsmokers.

Using Dr. Levin's methodology, the figures have been updated by the National Cancer Institute in their Smoking and Health Program. Tables 3:1 and 3:2 summarize some gruesome facts for the years 1965 to 1977. During this period, more than *four million deaths* are estimated to have occurred prematurely from tobacco usage!

Table 3:1. Estimated premature deaths attributable to tobacco usage.

Disease	Total Deaths 1977	Premature Deaths 1977	Premature Deaths 1965-1977
Diseases of the heart	755,250	226,600	2,871,700
Arteriosclerosis	33,000	10,900	314,900
Bronchitis/emphysema	20,350	17,300	142,700
Cancer of the trachea, bronchus and lung	89,000	77,450	776,300
Cancer of the oral cavity	8,450	5,900	68,800
Cancer of the larynx	3,350	1,650	19,500
Cancer of the bladder and kidney	17,100	5,350	60,600
Cancer of the esophagus	6,900	2,100	24,200
Cancer of the pancreas	19,800	6,950	82,100
TOTAL	953,200	354,200	4,360,800

Table 3:2. Annual premature deaths associated with tobacco use.*

Year	Number of Premature Deaths
1965	309,400
1966	318,700
1967	319,900
1968	333,000
1969	322,100
1970	333,600
1971	338,400
1972	346,500
1973	348,600
1974	344,400
1975	338,600
1976	343,400
1977	354,200

*1977 is the most recent year for which these figures are available, but it is safe to assume that the upward trend has continued.

The tobacco industry has used a second tactic to try to discredit statistics on excess deaths. Over the years, a number of different organizations have made estimates ranging from 125,000 to as high as 500,000. The tobacco people note this to suggest that all estimates are unreliable. The highest estimate, a guess made during the late '60s by Dr. David Horn was later revised downward by him as more specific data were accumulated. The simple fact is that today's estimates of excess deaths—well over 350,000 per year—are based upon valid analyses of massive amounts of data.

Ploy #12: "We are as old as America itself, and we kick in Big Bucks to the national economy."

Both of these claims are true. And the tobacco publications make sure we don't forget it, with ideas like these:
"Tobacco launched America."
"Tobacco: rooted deeply in America life, America's first commerce"
"When Christopher Columbus arrived in New York in 1492, he saw natives using tobacco."
"Tobacco is among the most heavily taxed consumer products, provides more than $7.0 billion annually in direct taxes to federal, state and local governments, helping to pay for such things as mass transportation and snow removal."
"Leaf is grown in 20 states on an estimated 276,000 farms. It is the sixth largest U.S. cash crop."

All of this may be interesting, but are tradition or economic facts more important than the hundreds of thousands of Americans killed each year by cigarettes? Simply because tobacco is as old as the country itself, should we continue to endure its carnage? The industry is heavily taxed (though these taxes may be less than the medical bills run up by smokers) and provides jobs for thousands of Americans. Are these contributions sufficient to balance 350,000 deaths per year and countless cases of human suffering, days lost from work, and property destruction from cigarette-induced fires?

Ploy #13: "Some of the studies have flaws—therefore, all of them probably do."

The tobacco team has a knack for finding a vulnerable part and kicking it repeatedly. The most vivid example is the Tobacco Institute's response to the recent so-called "Japaneses study" which suggested that nonsmoking wives of smokers had a higher rate of lung cancer than did nonsmoking wives of nonsmokers. The study's implications were indeed frightening: you could get cancer just by living with someone who smoked.

When this study was released, some of the more militant anti-smoking groups accepted it uncritically to add emphasis to their campaigns. But the scientific community looked at the study with great suspicion. It simply didn't seem plausible that nonsmokers inhaling diluted smoke through the nose could have cancer rates close to those of smokers inhaling concentrated smoke directly into the lungs.

As noted in Chapter 19, the design of the study was flawed, and it was soon contradicted in a study by Lawrence Garfinkle of the American Cancer Society. The Tobacco Institute responded with ads across the country depicting headlines saying "Scientists dispute findings of cancer risk to nonsmokers." Further research may or may not confirm the findings of the Japanese study, but the tobacco industry would like us to believe that if one study can be flawed, *all* evidence linking smoking and disease is questionable.

Ploy #14: "It's a free country, and those defending free enterprise and individual liberties are with us."

Surrounding themselves with political conservatives, American history books and the American flag, the tobacco interests talk about freedom and individual choice. But since researchers report that nine out of ten smokers wish they could stop, one might legitimately wonder whether smoking represents freedom or enslavement! Nor does the tobacco industry address the issue of freedom for nonsmokers who are burdened with the economic costs that cigarette smokers impose on the rest of our society.

4

In the Beginning, There Were No Cigarettes

> *For many centuries, tobacco use did not involve direct inhalation into the lungs. Only during the last century has tobacco smoke been* inhaled *on a regular and continuous basis by a sizable portion of our population.*

When most people refer to smoking, they immediately think of cigarettes. Pipes, cigars, chewing tobacco, and even some snuff are used today—but for the past few decades, particularly since World War I, cigarettes have been the dominant tobacco product.

It wasn't always this way. Indeed, with rare exceptions, until the approach of the 20th century, tobacco use had a relatively small impact upon human health. Before cigarettes became popular, other tobacco products had taken a toll; but a sharp distinction should be made between the pre- and post-cigarette eras. For many centuries, tobacco use did not involve direct inhalation into the lungs. Only during the last century has tobacco smoke been *inhaled* on a regular and continuous basis by a sizable portion of our population.

This chapter focuses on the pre-cigarette years.

A curious habit

Humans are the only animals who inhale smoke voluntarily into the mouth, nose and lungs. Smoking is a curious habit, one which you might think would be instinctively avoided. But this has not been the case. Since the beginning of recorded history, humans have smoked, sometimes in pursuit of relaxation, sometimes hoping to cure disease or promote health, sometimes in ritual. Smoking began long before tobacco was on the scene.

According to Susan Wagner's *Cigarette Country,* in ancient Greece, the priestesses at Delphi enveloped themselves in a cloud of smoke before delivering their prophecies. Herodotus described his practice of placing hemp seeds on hot stones after his evening meal to create an aroma he considered relaxing, even intoxicating. Hippocrates, Pliny and Galen prescribed inhalation of smoke—derived from burning coltsfoot, dried cow dung (particularly recommended for "melancholy") and other substances—for asthma and other diseases. The supposed therapeutic and relaxant effects of smoke from opium, marijuana and common herbs are a constant part of human history.

It has been established beyond reasonable doubt that tobacco was a purely "New World" plant, first used as a smoking substance in some remote time in the religious ceremonies of priests in the coastal regions of Central and South America. According to *Cigarette Country,* the oldest known evidence of use was found on a Mayan stonecarving at Palenque in the state of Chiapas, Mexico, adjacent to Guatemala—a bas-relief portraying a priest blowing smoke through a long tube. The Palenque slab dates back to the classic Mayan period, which would put it somewhere between 600 and 900 A.D. Native Americans were the first tobacco smokers; but it was their European conquerors who turned tobacco use into a global habit.

Turning a new leaf in the New World

The Indians had great respect—indeed a sense of honor—toward the tobacco leaf. They considered it a medical aid. They smoked it at religious occasions. They used it to combat weariness, pain and hunger. In 1492, when Christopher Columbus discovered the New World, his introduction to tobacco was an unexpected part of the deal. When he landed on San Salvador Island, the natives offered him food, drink, artifacts and a few handfuls of dried leaves. Columbus was not impressed with the leaves and threw them out. But he noted in his diary three days later:

> In the middle of the gulf between these two islands . . . I found a man alone in a canoe who was going from . . . Santa Maria to Fernandina. He had food and water and some dry leaves which must be a thing very much appreciated among them, because they had already brought me some of them as a present at San Salvador.

Later, on another island, Columbus and his men reported seeing people who "drank smoke." In the words of Bartholomio de Las Casas, who edited Columbus' journal:

> They did wrap the tobacco in a certain leaf in the manner of a musket formed of paper . . . having lighted one end of it . . . they sucked, absorbed or received that smoke inside with their breath.

Other sources suggest that the leaves described by Bartholomio were those of maize or corn.

Within 40 years after the discovery of America, Spaniards were cultivating the crop commercially in the West Indies. And thus began an era dominated by such colorful figures as Jean Nicot, the French ambassador to Portugal who wrote to a friend in 1560 that an American herb he had acquired had marvelous curative powers. As a result, consumption of tobacco gained popularity and Nicot earned a place for himself in the history books. Indeed, his name became the basis for the scientific term for tobacco: *Nicotiana*.

Word about tobacco spread quickly. A detailed account of the herb tobacco was written in 1577 under the title, *Joyfullnewes oute of the newfounde worlds*. This volume notes:

> In like sat the rest of the Indians for their pastime, doe take the smoke of the Tobacco, to make themselves drunke withall, and to see the visions, and things that represent unto them that wherein they do delight.

The history books suggest that it was Sir John Hawkins, in about 1565, who first brought tobacco to England. But Sir Walter Raleigh became tobacco's leading publicist. He even convinced Queen Elizabeth I to try it, but it made her nauseated. Legend has it that when Sir Walter's manservant first found his master smoking, he thought he was on fire and doused him with beer. But despite these minor mishaps, Sir Walter did manage to popularize smoking in England during the late 1500s with at least the tacit approval of the Queen. By 1603 English poet John Manston classified tobacco among the wonders of the world: "music, tobacco, sack (wine) and sleep; the tide of sorrow backward keep."

Queen Elizabeth's successor to the throne, James I, despised tobacco. He considered its use unhealthy, unholy and altogether offensive to civilized society. He began to tax tobacco heavily in an effort to limit its use. King James—who could be considered the patron saint of the anti-tobacco movement—is best remembered for his oft-quoted *Counterblaste to Tobacco,* which he had published anonymously in 1604. He described tobacco as:

> A custome lothesome to the eye, hatefull to the Nose, harmfull to the braine, dangerous to the Lungs, and in the black stinking fume thereof neerest resembling the horrible Stigian smoke of the pit that is bottomelesse.

The year following publication of his *Counterblaste,* King James organized at Oxford the first public debate on the effects of tobacco. He used props to get his message across: black brains and black viscera, allegedly obtained from the bodies of smokers. Later he lamented:

> Herein is not only a great vanity, but a great contempt of God's good gifts, that of the sweetness of man's breath, being a good gift of God, should be willfully corrupted by this stinking smoke.

James' *Counterblaste* might be considered one of the first formal and widely circulated critiques of tobacco. Many more followed during the 18th and 19th centuries. What is of interest here is that today's defenders of cigarettes—a category limited almost exclusively to those who sell them—like to claim that today's criticisms of smoking based on scientific studies are equivalent to the opinion and speculations from the days of King James.

In an article reprinted from *American Heritage* and distributed by the Tobacco Institute, author Gordon L. Dillow refers to James' writing as "the world's first antismoking tract"—an "opening shot in the conflict" that would eventually lead to present-day concerns. William F. Dwyer, Vice President of the Tobacco Institute, wrote in the late 1970s:

> The tobacco controversy is ages old. Several sociologists suggest that the core of this controversy is an ineluctable part of human nature in that any practice or product which provides pleasure for some will evoke outrage in others.

In his book on cigarette advertising, *Selling Death,* Thomas Whiteside quotes Philip Morris executive James C. Boling's perspective on what he calls a "health scare" as follows:

> I remember a speaker last year at the 350th anniversary celebration of America's first tobacco crop at Jamestown telling of the trials that the tobacco industry had at that time. They had a health scare at the inception of the industry in America. And the scare goes farther back than that.... We've had these trials from time to time, and each time the industry has come through stronger, because these people have demonstrated conclusively that they want to smoke.

These comparisons are nonsense! Today's criticism is based upon incontrovertible scientific evidence of harm.

King James, who found tobacco champion Sir Walter Raleigh's politics about as unsavory as his smoking, put up considerable resistance to the spread of tobacco use. James raised taxes on tobacco 40-fold, limited Virginia production to 100 pounds per year, and forbade tobacco growing in England, saying that the growth of food for the colonies would be jeopardized. He levied fines and sometimes ordered corporal punishment for those who did not pay his tobacco taxes. But eventually even James gave in. He allowed London, and later Bristol, to have importing privileges but required retailers to be licensed. In the art of royal compromise, he promoted a "cigarettiquette," leading to the custom of men requesting permission before they smoked, and the well known after-dinner sanction, "Gentlemen, you may smoke."

During James' age, the poor and the rich smoked differently. Given the high cost of tobacco, the economically less fortunate adulterated the tobacco leaf with all sorts of ingredients. They made their own pipes, sometimes from walnut shells. The rich surrounded themselves with all sorts of smoking paraphernalia, including exquisitely carved tobacco boxes, pipe cleaners, scrapers, silver pipes and other ways of enhancing the pleasure from this new fashionable habit.

Money for the Colonies

Despite James' efforts, tobacco use spread in England, then quickly throughout the world. As one pipe lover put it toward the end of the 19th century, "Prince and peasant alike yielded to its mild but captivating sway."

One key element in the increasing use of the tobacco leaf was the struggling colonies' need for an export product. In the early 1600s, when people began settling in Virginia, they assumed that they would make their fortunes from the New World's precious metals, lumber, furs and fish—but it was soon discovered that tobacco brought in substantially more money than any of these commodities.

John Rolfe, the young Englishman who eventually married Pocahontas, settled in Jamestown in 1610 and made it the home of tobacco country. (How ironic that the town is named after tobacco's most vocal critic!) Rolfe knew that Virginia needed an export staple to give it economic stability. It is said that on one occasion when John Rolfe returned to Jamestown after a voyage, he discovered tobacco literally growing in the streets. Noting the growing demand for tobacco in England and elsewhere, Rolfe soon realized that exporting this product was the one way to ensure survival of the new community. He imported some seed of *Nicotiana tabacum*, which Spanish colonists were producing in the Caribbean Islands, a type of tobacco which was particularly acceptable to English smokers. As one historian put it, "Never was a marriage of soil and seed more fruitful." Jamestown soon became a tobacco boomtown (Rolfe's first shipment went to England in 1613)—and farmers all over the southeastern colonies quickly realized that in Virginia, growing tobacco was many times more profitable than growing corn.

The import duties which tobacco provided grew impressively, as did the demand for the product in England. In 1615, Jamestown exported 2,300 pounds of tobacco; three years later 20,000 pounds and in 1620, 40,000 pounds. Suddenly, James I's tobacco taxes became lucrative.

In 1622, some of the directors of the Virginia Company still looked upon smoking as a temporary fad. Tobacco, they said, was a "deceivable weed which served neither for necessity nor for ornament of the life of man but was founded only upon a humour which must soon vanish into smoke." They could not have been much farther off in their predictions.

Deterrent efforts

While smoking was becoming accepted or at least tolerated in England and the colonies during the 1600s, some countries went to considerable length to try to stop the practice.

In 1606, Philip III of Spain issued a decree restricting the cultivation of tobacco. In 1610 Japan issued orders against both smokers and planters. Things really got tough in Russia where the patriarch placed smoking and snuffing in the category of deadly sins! The Czar, in 1634, issued a ukase against the use of tobacco, stating that first offenders—smokers or vendors—would have their noses slit. In a number of cases, repeat offenders were sentenced to death. In Berne, smoking was considered as sinful as adultery and was punished accordingly. The Pope ordered that persons taking tobacco or snuff into a Roman Catholic Church be excommunicated—cigar and pipe smoke were competing with the aroma of incense, and some monks were coughing during the solemn chants.

Whippings, beheadings, nose slittings and other extreme measures, however, were unable to stem the tide of tobacco acceptance. Murad IV of Turkey is said to have gone to extremes in his antismoking crusade. He reportedly roamed the streets of 17th century Istanbul in disguise trying to get vendors to sell him tobacco. If they did—an act against his official policy—he would behead them on the spot, leaving the body in the street as a grisly warning to other would-be lawbreakers. (Nonetheless, three centuries later, a Turkish tobacco cigarette was introduced called Murads!)

American efforts to dissuade people from tobacco use were relatively mild. In the 1630s, the Massachusetts colony banned tobacco sales and public smoking. In the 1640s, Connecticut also banned public smoking and required smokers to obtain a permit. These laws, however, were generally ignored and soon were repealed.

Tobacco and health: the 17th century view

During the 17th century, tobacco was regarded as an all-purpose healer—a "magical herbe." In 1602, in *Natural and Artificial Directions for Health,* Sir William Vaughn stated, "Cane tobacco well dried, and taken in a silver pipe, fasting in the morning, cureth the megrim, the toothache, obstructions proceeding of cold, and helpeth the fits of the mother (hysteria)."

During the plague year of 1603, a Tom Rogers reported that Eton schoolboys were required to "smoake in the school every morning." Mr. Rogers, ironically, also reported that he never received a harsher whipping than he did for *not* smoking one day at school. During the Great Plague of 1665, those who believed that tobacco conferred immunity advised that even nonsmokers chew or smoke to protect themselves. Mothers furnished schoolboys with filled pipes, and schoolmasters taught children how to smoke.

In the early 1600s, Robert Burton wrote in *Anatomy of Melancholy*: "Tobacco, divine, rare, superexcellent tobacco, which goes far beyond all the panaceas, potable gold and philosopher's stones, a sovereign remedy to all diseases." And in 1614, William Barclay, Doctor of Physicke who apparently agreed, wrote: "It prepareth the stomache for meat; it maketh a clean voice; it maketh a sweet breathe."

During this era, however, there were occasional warnings. In 1606, a Doctor of Physic advised that tobacco was not safe for youth, shortened life, bred many diseases, including melancholy, "hurteth the mind," and generally was "ill for the smokers tissue." In 1689 the Medical School of Paris officially announced its view that smoking shortened life. But overall, tobacco emerged from the 17th century with a good reputation.

The golden token

Through the 1700s, tobacco secured its grip on the economic structure of the southeastern part of what would soon become the United States. Indeed, tobacco's designation, "the golden token," said it all. Tobacco was a fully negotiable commodity that could be exchanged for European products and for cash. The rich soils of the James River region yielded 1,660 pounds of tobacco per acre on the best land, never getting less than 500 pounds per acre. Indeed, if tobacco farmers had any problem in the 18th century, it was one of overproduction.

During this period, smoking clubs became the rage. Paintings portrayed beautiful women gracefully smoking long-stemmed pipes—an activity which was not uncommon. Even the English language documented the respect and influence of tobacco. To "put out someone's pipes" meant dressing the person down, putting him in his place. "Put that in your pipe and smoke it" still conveys the message, "Face up to this or else"; "pipe dreams" are baseless hopes, reveries as transient as the smoke itself. Charles Kingsley wrote:

> Tobacco, a lone man's companion, a bachelor's friend, a hungry man's food, a sad man's comfort, a wakeful man's sleep, and a chilly man's fire..there is no herbe like it under the canopy of heaven.

Tobacco and health: the 18th century view

During the 1700s, some interesting and prophetic observations were made by physicians. In 1713, Italian physician Bernardino Ramazzini, a pioneer in occupational medicine, wrote in *De Morbis Artificum Diatriba* that tobacco workers, although they loved their jobs and had no intention of giving them up, did pay a price because headaches and stomach disorders were widely known to be side effects of inhaling tobacco dust. "The sweet smell of gain," Ramazzini commented, "makes the smell of tobacco less perceptible and less offensive to the workers. . . . This vice will always be condemned and always clung to." Ramazzini, obviously negative about the tobacco habit, lamented that "women as well as men and even children" were using it, and its purchase was "reckoned among the daily expenses of a family."

In 1761, English physician Dr. John Hill made the first clinical report documenting tobacco as a cause of cancer. Dr. Hill presented his findings in a document entitled *Cautions Against the Immoderate Use of Snuff*. He reported two cases of cancer of the nose, both victims having "polypusses," which he believed to be malignant with "very bad consequences."

> This unfortunate gentlemen, after a long and immoderate use of snuff, perceived that he breathed with difficulty through one of his nostrils; the complaint gradually encreased 'till he perceived a swelling within. . . . It grew slowly, till in the end, it filled up that whole nostril, and swelled the nose so as to obstruct the breathing . . . he found it necessary to then apply for assistance. The swelling was quite black and it adhered by a broad base, so that it was impossible to attempt the getting it away . . . and the consequences was a discharge of a thick sharp humor with dreadful pain, and all the frightful symptoms of cancer . . . and he seemed without hope when I last saw him.

His second case was also a victim of snuff:

> The person was a lady of a sober and virtuous life . . . she had long been accustomed to snuff and took it in a very great quantity . . . she felt a strange soreness in the upper part of her nostril . . . after a little time, came on a discharge of a very offensive matter; not in any quantity but of an intolerable smell, and was more so to her, as she was naturally a person of great delicacy. The discharge encreased, and it soon became necessary for her to leave off snuff. A surgeon was employed, but to very little purpose.

In 1798, Dr. Benjamin Rush, a signer of the Declaration of Independence, wrote an essay called "Observations Upon the Influence of Habitual Use of Tobacco Upon Health, Morals and Property." His primary objection to tobacco was that he felt it caused disastrous effects on the stomach, nerves and oral cavity. He believed its use was a filthy and expensive habit. He also suggested a direct cause-and-effect relationship between tobacco use and drunkenness, overstating what may have been genuine clinical observations.

On the verge of the cigarette

The 1800s were an exciting period for tobacco in the United States. During this century the use of snuff gave way to continued use of pipes and increased popularity of cigars. Toward its end, a new champion emerged: the cigarette.

One sign of the times—an indication that tobacco use was becoming an important social activity—was that rail carriages for smoking were introduced in England in 1846. One cartoon even depicted a man complaining because a fellow traveler wasn't smoking. Also during this decade, the first organized and strong anti-cigarette movement was born and, curiously enough, a substantial amount of clinical evidence accumulated concerning the fact that tobacco use caused cancer and other diseases.

Although the wives of Andrew Jackson and Zachary Taylor smoked pipes in the White House, as the 19th century wore on, it became increasingly unpopular for the "delicate sex" to smoke. By the time the cigarette was introduced, it was clearly considered unacceptable for women to smoke.

The number of cigarette factories in America increased enormously, especially in Virginia and North Carolina. In 1840 there were some 119 such factories; by 1860 the number totaled 328.

The birth of the anti-tobacco movement

There had always been some opposition to tobacco—that it was distasteful, annoying, hazardous to health or simply immoral, but during the 1800s, sentiment against tobacco became more formal. A number of prominent people aligned themselves to combat what they saw as the evil effects of the weed. In the majority of these situations, however, the objections to tobacco were based on personal opinion rather than scientific evidence. Indeed, much of the anti-smoking effort seemed to be based on the Puritan ethic, or in some cases, the 19th century ambivalence about sex:

> No man can be virtuous as a companion who eats tobacco for, although he may not violate the seventh commandment, yet the feverish state of the system which it produced necessarily causes a craving and lustful experience of amativeness. . . . You who would be pure in your love instinct, cast this sensualizing fire from you.

In 1859, Reverend George Trask wrote a widely circulated book (with a typical 19th century title): *Thoughts and Stories for American Lads: Uncle Toby's Anti-Tobacco Advice to His Nephew Billy Bruce.* Citing some dubious statistics, he stated, "Physicians tell us that twenty thousand or more in our own land are killed (by tobacco) every year." Sounding vaguely like the 1982 Surgeon General's report, he added, "German physicians tell us that of deaths of men between the ages of 18 and 25, one-half originate from this source."

Just prior to the Civil War, a group of physicians, educators and men of the church—along with the great P.T. Barnum—joined forces to fight the tobacco habit. They were soon joined by John Hartwell Cocke, co-founder of the University of Virginia, a big name in the anti-slavery and anti-liquor movements of previous and current years. Mr. Cocke wrote a piece entitled *Tobacco: The Bane of Virginia Husbandry.*

Horace Greeley, publisher of the *New York Tribune,* became part of the anti-smoking movement. Calling tobacco "narcotic sensualism," he defined a cigar as "a fire at one end and a fool at the other."

Nineteenth century views

It is fair to say that during the 19th century, there were three levels of evidence suggesting that tobacco was hazardous to health. One was based on informal clinical observations; another on more formal investigations; and the third on well-publicized accounts of individuals who had died from the apparent effects of tobacco use, generally cigar smoking.

In 1836, Samuel Green blasted tobacco in the *New England Almanack and Farmer's Friend*:

> Smoking—That tobacco may kill insects on shrubs and that one stench may overpower another is possible enough; *but that thousands and tens of thousands die of diseases of the lungs generally brought on by tobacco smoking, is a fact as well known in the whole history of disease.* How is it possible to be othewise? *Tobacco is a poison.* A man will die of an infusion of tobacco as of a shot through the head. *Can inhaling this powerful narcotic be good for man?* Its operation is to produce a sensation of giddiness and drowsiness—is it good to be within the next step of perpetual drunkenness? It inflames the mouth and requires a perpetual flow of saliva, a fluid known to be among the most important to the whole economy of digestion; it irritates the eyes, corrupts the breath and causes thirst. No doubt the human frame may become so far accustomed to this drain, that the smoker may go on from year to year making himself a nuisance to society, yet there can be no doubt whatever, that the custom is as deleterious in general as it is filthy.

Green's evaluation of tobacco was anecdotal, not based on scientific data. But data were gradually accumulating. A Boston surgeon wrote in 1849:

> For more than 20 years back I have been in the habit of inquiring of patients who come to me with cancers . . . of the gums, tongue and lip . . . whether they use tobacco. . . . When, as it usually is the case, one side of the tongue is affected with . . . cancer, the tobacco has habitually remained in contact with that part.

During the same year, Dr. Joel Shew published a book entitled *Tobacco: Its History, Nature and Effects on the Body and Mind* and asked a prophetic question: "I believe cancers . . . and tumors in and about the mouth will be found much more common among men than women. Since the former use tobacco much more generally than the latter, may not this be the cause?" He carefully catalogued (in a style described in a *Tobacco Institute* reprint as "often in repellent detail") some 87 maladies which he attributed to tobacco use, including insanity, cancer and hemorrhoids.

In 1851, Sir James Paget saw a patient with leukoplakia (then known as "Smoker's Patch") near the middle of the upper surface of the tongue where he always rested the end of his pipe, and "told him he certainly would have cancer of the tongue if he went on smoking."

In 1859, Bouisson, a French physician reported a remarkably complete clinical study of 68 patients with cancer of the oral cavity. Two-thirds of the cases were lip cancers; the others were cancers of the tongue, mouth, internal surfaces of the cheek, tonsil and gums. Identifying the smoking habits of 67 of these patients, Bouisson found that 66 of them smoked tobacco and the other one chewed it. He also determined that cancer of the lip ordinarily occurred at the spot where the pipe or cigar was held!

In 1856 and 1857, the prestigious British medical journal, *Lancet* featured a series entitled "The Great Tobacco Question" in which 50 doctors expressed their views. A Dr. Hodgkin associated tobacco with an increase in crime. He also claimed that the use of tobacco, by drying the stomach, caused a craving for alcohol, thus leading the smoker into the dangerous and immoral cult of Bacchus. Indeed, Dr. Hodgkin thought the word tobacco was derived from "To Baccho"! A Dr. Solly associated it with nervous paralysis and a loss of intellectual capacity. And a Dr. Schneider claimed, "So frequently is vision impaired by the constant use of tobacco that spectacles may be said to be part and parcel of a German, as a hat is to an Englishman . . . Americans wear themselves out by the use of tobacco."

But, at the end of the series the medical journal itself gave a relatively clean bill of health to tobacco:

> The use of tobacco is widely spread, more widely than any one custom, form of worship, or religious belief, and that therefore it must have some good or at least pleasurable effects; that, if its evil effects were so dreadful as stated the human race would have ceased to exist.

In the March 14, 1857 issue of *The Lancet,* however, there appeared a commentary that remains relevant today:

> Tobacco is said to act on the mind by producing inactivity thereof; inability to think; drowsiness; irritability. . . . On the respiratory organs, it acts by causing consumption, haemoptysis [coughing up of blood] and inflammatory condition of the mucous membrane of the larynx, trachea and bronchae, ulceration of the larynx; short irritable cough, hurried breathing. The circulating organs are affected by irritable heart circulation.

Some public discussion followed the death of General Ulysses S. Grant, an inveterate cigar smoker who died of throat cancer in 1885. In the public's mind, his cigar-smoking habit was associated with his military prowess. One of his most cherished gifts from his admirers after his victory over the South was a collection of 11,000 cigars. Medical reports indicate that in his last months, General Grant's deterioration and agony were such that he had to sleep sitting up to relieve the pressure and pain of his spreading malignancy. Grant's detailed obituary in *The New York Times* of July 24, 1885, makes it clear that his doctors thought his disease had been caused by smoking:

> On June 2, 1884, while eating his lunch at Long Branch, the General, as he tasted some fruit, felt a lump in the roof of his mouth and found that swallowing was painful. The lump grew more troublesome day by day. The General was an'inveterate smoker, and his cigar on the battlefield has become as much a matter of history as the story itself. To give up a life-long habit, which had been so confirmed as this, was no easy task and the physicians, recognizing this fact, confined their advice to requesting him to limit his indulgence in tobacco. They recommended him to confine his smoking to three cigars a day.

Few Americans paid much attention to these reports. Medical communication was slow and tedious, and medical opinions made their way to public consumption slowly if at all. Compared with modern times, there were relatively few smokers, and most of those who did smoke did so only occasionally. Moreover, cancer was less well-known and less feared than it is now. If tobacco use had been associated with tuberculosis or one of the other diseases dreaded in the 19th century, people might have been more concerned.

In 1879, there were some 40 million Americans, slightly under half of whom were men. During that year, the nations' factories and importers made available more than 1.2 billion cigars—about 60 cigars per male citizen (infants included). Another 100 million pounds of other tobacco material—snuff, pipe and chewing tobacco—were also used. Tobacco was most definitely a growth industry, with production close to doubling during the next year. But its use was still somewhat cumbersome—requiring stuffing and re-stuffing a pipe, or finding a place to light a cigar.

The arrival of the cigarette and the portable match changed all of that—enabling anyone, anywhere, to light up, conveniently and frequently—and (unlike pipe and cigar smoke which is far more alkaline) provided the additional "advantage" of a smoke "mild" enough to inhale directly into the lungs.

5

A Triumph of American Ingenuity

"I vow and believe that the cigarette has been one of the greatest creature-comforts of my life—a fine companion, a gentle stimulant, an amiable anodyne, a cementer of friendship."—William Makepeace Thackeray

One might consider the birthday of the American cigarette to be April 30, 1884, the day a team of clever, enterprising men worked out the final hitch to the cigarette manufacturing machine. This device increased the supply of cigarettes enormously and made them the focus of the most lucrative advertising, promotion and sales effort in history.

Cigarette-like products had been in limited use for centuries. Apparently cigarettes had originated in Central and South America where Indians puffed on reeds filled with tobacco. During the 1800s, the primitive cigarette began to gain in popularity in Spain where the reed was replaced by a paper sheath that held shredded tobacco, and the proverb of the time was "A paper cigarette, a glass of fresh water, and the kiss of a pretty girl, will sustain a man for a day without eating." Georges Bizet's opera *Carmen* featured a heroine who worked in a cigarette factory in Seville around 1829.

The word "cigarette" is of French origin, and the newly formed weed did pass through that country as it did through others in Europe. In 1848 the French made cigarette sales a government monopoly.

History books indicate that British soldiers were introduced to cigarettes during the Crimean War of 1853-56 and took this strange curiosity home with them. The first British cigarette factory was opened in 1856 by Robert Gloag who smoked a smoke-cured form of Turkish tobacco. In the late 1850s, London tobacconist, Philip Morris, went into "Hand Made" to produce Turkish cigarettes. By 1890, cigarettes were popular on the Continent. In that year Oscar Wilde's *The Picture of Dorian Gray* included the line, "Basil, I cannot allow you to smoke cigars. You must have a cigarette. A cigarette is the perfect type of a perfect pleasure.... What more can one want?"

Image problems

British acceptance of the "little smokes" helped pave the way for American acceptance. American visitors to Britain brought cigarettes home, but they hardly took America by storm. Indeed, they quickly developed a bad reputation. Rumors spread that their paper was saturated with opium and arsenic, and that their tobacco was derived from cigar butts left in the gutters by derelicts. More revolting was the widely circulated report that cigarette factory workers urinated on the tobacco to give it "bite."

Cigarettes had another image problem: they were considered effeminate for men yet too masculine for women. The opera *Carmen*, which opened in Paris in 1874, associated cigarette use with being low-class. Some of the names of the newly introduced cigarettes ("Opera Puffs" and "Pearl's Pets"), along with the "ette" suffix which gave it a diminutive air, made this problem even more prominent. Thus, men who adopted this novelty had their masculine reputation on the line. "The cigarette is designed for boys and women" *The New York Times* declared in 1884. The *Times* also added an opinion on the cigarette when it editorialized, "The decadence of Spain began with the Spaniards adopting cigarettes, and if this pernicious practice remains among adult Americans, the ruin of this Republic is close at hand."

For many years only the most daring woman would be seen smoking a cigarette, even in cosmopolitan New York. Her morality and image were at stake. Word had it that, "When one hears of a sly cigarette between feminine lips at a croquet party, there is no more to be said!"

Production begins to skyrocket

Statistics on U.S. cigarette production date from 1880. That year, with just over 50 million people, the United States consumed 1.3 billion cigarettes, 500 million made here and the rest imported. The cigarette constituency was small, made up of immigrants, eastern city "dandies" and poor people. The tobacco trade was still dominated by chewing tobacco, cigars and pipe tobacco, but cigarettes were definitely on their way up. Small hand-rolling cigarette factories opened up and promotion got started. Cigarette cards initially meant to serve as package stiffeners were soon picturing actresses. The moralists took note, protesting loudly against the use of cigarette cards that showed buxom beauties with "luscious legs." By the late 1880s, cigarette-smoking boys were a common sight on any urban street corner. The give-away cards in the cigarettes were very much in demand by the younger set who traded and gambled them like the bubblegum baseball cards of the 1950s.

The one major obstacle to growth of the cigarette habit in the United States in the 1870s and 1880s was the fact that the product required hand rolling—with much human effort. Because the most skillful and dedicated roller could produce only four cigarettes a minute, there were not enough cigarettes available to meet the growing demand. Day after day, cigarettes were rolled by hand on marble slabs, sometimes with hired "readers" trying to alleviate the roller's boredom.

All that changed—and the future of the cigarette suddenly became brilliant—when James Albert Bonsack invented the first practical cigarette-making machine which he patented in 1881. Bonsack had been working on the rolling machine since his mid-teens. By age 22 he had a device which, when perfected, could produce 120,000 cigarettes a day, the equivalent of 40 expert rollers working 12½ hours. Because Bonsack's machine had some technical problems (the flow of shredded tobacco toward the rollers often stalled, slowing production), he went into business with W. Duke and Sons Company, accepting modest royalties.

Washington Duke, whose son James Buchanan Duke later formed the American Tobacco Company, had a small tobacco farm near Durham, North Carolina. William Bennett, writing in *Harvard Magazine*, noted that, "After Appomattox, Washington Duke . . . was released from Libby prison. His only remaining assets were two blind mules, a farm in North Carolina and a barn full of tobacco leaf purchased before he went to war."

In a little log cabin farm near Durham, Duke and his four sons raised "baccy" for a living. He had developed a small successful business, but meeting Bonsack was his big break. The four biggest tobacco companies had rejected Bonsack's machine because it was mechanically unreliable, but Duke saw that improvements could be made. Working with Duke's best mechanic, William O'Brien, Bonsack perfected the machine. On its final test, April 30, 1884, each machine successfully operated for the full work day.

In 1881 Duke's factory produced 9.8 million cigarettes, 1½ percent of the total market. But five years later, W. Duke and Sons were able to manufacture 744 million cigarettes, more than the national total in 1883. By 1890, Duke's competitors, who themselves had now become mechanized, joined forces with him to establish the American Tobacco Company. By the turn of the century, 9 out of every 10 cigarettes carried the Duke label. But a few months after the American Tobacco Company was formed, the state of North Carolina started an antitrust suit against it— and other such litigation followed. In May 1911, the American Tobacco Company was dissolved by order of the Supreme Court, to be succeeded by four large firms—Liggett and Myers, Reynolds, Lorillard, and American—plus many smaller ones.

The market expands

With production problems behind him—thanks to the Bonsack machine and years of experience in raising tobacco—Duke (who, incidentally, never smoked his or anyone else's cigarettes) could turn to a new and very important phase: promoting his cigarettes, making them attractive and desirable commodities, and generally getting people to take up the habit. A 1933 *Forbes* article noted:

> In 1884 the Dukes came to New York, Southern "drawl" and all, and five years later astonished the financial powers by effecting a consolidation of five of the largest tobacco interests in the country. Then they proceeded to lay down an intensified advertising barrage. Duke's advertising expenses were enormous. . . . He threw hundreds of thousands of dollars into the battle.

Duke examined the packaging of popular brands and felt there was room for improvement. The traditional packaging, much like that used today, was soft, offering no protection to the cigarettes, which could be easily crushed. Duke came up with a slide-and-shell hard back, more expensive to produce, which added a touch of elegance and class to the cigarette.

Duke was a good businessman. He bragged on his packaging that, "These cigarettes are manufactured on the Bonsack Cigarette Machine"—and he priced his products right. In 1885 he sold a brand called Pin Head at 10 for 10¢. He hired innovative people to generate press coverage for his product. For example, Duke engaged the services of Edward Featherston Small, an ambitious businessman, giving him a nearly unlimited budget for the purpose of promoting sales. When the traditional means of advertising didn't work in Atlanta, Small threw out the Wooden Indian carvings and posted pictures of Madame Rhea, a popular French actress sporting a package of Duke cigarettes. When he had trouble breaking into the St. Louis market, he hired an alluring woman, Mrs. Leonard, as a salesperson. With a touch of genius, Small called a press conference to highlight a woman on the job—and generated front page publicity for Mrs. Leonard, and more important, for Duke's cigarettes.

Cigarettes began to appeal to a more prosperous and refined public, one which did not use chewing tobacco, pipes or cigars. By 1885, a variety of brands with enticing names were circulating: Vanity Fair, Fragrant Vanity Fair, Cloth of Gold, Three Kings, Old Gold, Bon Ton, Napoleons, Dubec, Melrose, and Golden Age.

The cigarette machine was obviously a real boon for the industry. But Duke also took full advantage of another invention—or rather a technological improvement—which had an equally dramatic effect on increasing sales: the portable match.

Matches that existed before the turn of the century were downright dangerous. Many people considered their fumes more harmful than the lighted tobacco could ever be. Rather than use the old-fashioned matches, cigar smokers generally lit up with embers from the hearth or a taper from an oil or gas lamp. Smokers of the 1890s would no more choose to carry matches than they would choose to carry an explosive on their person. Without a handy match, cigarettes and other smoking material could still be ignited, but not conveniently, and the lack of convenience cut down on the number of cigarettes smoked. As Robert Sobel notes in *They Satisfy*:

> What matches did was to alter the *way* cigarettes were smoked, encouraging their consumption during odd moments of the day; in effect, transforming their use from a thoughtful exercise into an almost unconscious habit.

Sobel points out that cigarette users doubled their consumption of cigarettes when safe matches became available. Soon after the turn of the century, they were being distributed free, in book form, with the purchase of cigarettes. Duke probably knew that his Bonsack machine and safe matches were "a match" for anyone—including the anti-smoking crusaders who were now on the scene in large numbers—and that the great American tobacco age was about to begin. The growth of the American cigarette industry was phenomenal. In 1889, only five years after the industry was mechanized, the United States produced 2.5 billion cigarettes, at least 40 percent of them being Duke's.

Vocal opposition

Tobacco products of all kinds had always gotten some bad press. But just before the turn of the century, the anti-smoking crusade took on a formal structure. Perhaps this was a response to the stepped-up promotions aimed at hooking new smokers, particularly young ones. "There is no question that demands more public attention than the prevailing methods of cigarette manufacturers to foster and stimulate smoking among children," one irate New Yorker wrote in 1888, sounding an alarm which is still relevant today.

The anti-tobacco forces were frequently strident, hysterical, irrational, emotional—and committed with an almost religious fervor to eliminating cigarettes. (Curiously, they ignored the other forms of tobacco use.) Their pleas were moralistic and similar to the growing sentiment against alcoholic beverages. Lacking today's clear-cut statistics on cancer, heart disease and emphysema, they conjured up medical "evidence," claiming, for example, that cigarette smoking caused color blindness, weak eyesight, baldness, stunted growth, insanity, sterility, impotence, sexual promiscuity, drunkenness, a tendency to turn to crime, and even immediate death. The following case, reported in an 1890 edition of *The New York Times* is typical of charges made at the time:

> New Jersey—The death of eight-year-old Willie Major, a farmer's son, from excessive cigarette smoking is reported from Bound Brook. The boy had, for over three years, been a victim to the habit. He would stay away from home several days at a time, eat nothing but the herbs and berries of the neighborhood and smoke constantly. Sunday he became ill and delirious. He died Tuesday in frightful convulsions.

At least as important as the alleged physical deterioration caused by cigarettes was the claimed impact of cigarettes on character. New York school commissioner Charles Hubbell wrote in 1893:

> Many and many a bright lad has had his will power weakened, his moral principle sapped, his nervous system wrecked, and his whole life spoiled before he is seventeen years old by the detestable cigarette. The "cigarette fiend" in time becomes a liar and a thief. He will commit petty thefts to get money to feed his insatiable appetite for nicotine. He lies to his parents, his teachers and his best friends. He neglects his studies and, narcotized by nicotine, sits at his desk half stupefied, his desire for work, his ambition, dulled, if not dead.

Lucy's cause

Lucy Page Gaston was born in 1860 into a midwestern family heavily involved in reform movements, including abolition and temperance. All of the Gastons were upright citizens, nonsmokers, nondrinkers and eager to make the rest of the world as holy and wholesome as they considered themselves. Lucy was probably the most successful anti-smoking crusader in history. Were she alive today and armed with 20 years of Surgeon Generals' Reports, the tobacco empire might be quivering with fear.

Lucy became committed to what she called "The Cause." Believing that the trained eye could spot smokers at a glance, she created the concept of the "cigarette face." Actually, there may have been something to her theory. Cigarette smokers often have prematurely wrinkled skin, saggy, baggy eyes and a telltale cough!

Miss Gaston had a jumble of pseudoscientific jargon to explain why cigarettes were harmful. There were 20 drugs in those coffin nails, she would say, and nicotine was the least of the problems. She was most concerned about something she termed furfural, produced, she claimed, by the burning of glycerine. Miss Gaston used her vivid imagination to get her message across. She organized groups of children to sing songs and carry anti-cigarette banners. She allied herself with public figures who opposed the use of cigarettes, such as Thomas Edison, boxing champion John L. Sullivan, Henry Ford (who said, "If you will study the history of almost any criminal, you will find that he is an inveterate cigarette smoker"), and author Elbert Hubbard (who said, "Never advance the pay of a cigarette smoker—never promote him—never depend upon him to carry a roll to Garcia unless you do not care for Garcia and are willing to lose the roll").

This alumna of the Women's Christian Temperance Union left the tobacco men shaking in their North Carolina boots. She searched newspapers for items on cigarettes, and after polishing them up a bit, reprinted them in a newspaper she started in Harvey, Illinois. One which she claimed came from the *Denver Post* read "Daffy, John Jones age 19, is very sick and at times acts very queer; caused by the excessive use of cigarettes."

The first big move to legislate against cigarettes began in earnest in the 1890s. Lucy Gaston and her comrades were having an impact. New York was the first to act, making it a misdemeanor for any person "actually or apparently under 16 years of age" to smoke in public. In 1897, a federal law banned the use of coupons in cigarette packaging. By the turn of the century New Hampshire made it illegal for any person or corporation to make, sell or keep any form of cigarette. In 1892, Congress was overwhelmed with petitions from WCTU groups stating that cigarettes were "causing insanity and death to thousands" of American youth and demanding that the offending products be banned. Cigar manufacturers whose business was obviously jeopardized by the growing popularity of the cigarette, may have supported this campaign, or at least did not offer much resistance to it. Cigarette production peaked at 4.9 billion in 1897 and then started to decline. By 1908, fewer than 3.5 billion were being produced.

But cigarettes had a firm grip on a small but growing number of American men. The new smokes had the "advantage" of being mild enough to inhale, and fit in perfectly with the newly quickened pace of life that went with increased industrialization and the rapid growth of cities. Pipes and cigars remained for the use of the dwindling number of men of leisure, those who had time to sip brandy, chat with friends and study great literature. Cigarettes were for the busy modern man on his way up.

6
When Camels Became Kings

Tobacco is a dirty weed, I like it.
It satisfies no moral need. I like it.
It makes you thin, it makes you lean
It takes the hair right off your bean
Its the worse darn stuff I've ever seen.
I like it.—Penn State Froth, *1915*

The period from the turn of the century to the end of World War I contained both spectacular triumph and bitter conflict for the cigarette in America.

Smoking was "in," and the great thinkers of the day wrote about it. Freud, a dedicated (20-a-day) cigar smoker who died of cancer of the mouth and jaw, related smoking to oral needs which begin with thumb-sucking in children: "It is not every child who sucks in this way.... If that significance persists, these same children ... will have a powerful motive for drinking and smoking."

Popular music reflected the growing acceptance of the cigarette. Albert Marrin, the Yeshiva University historian who writes for the *Tobacco Observer,* has chronicled how the new style of smoking inspired many songwriters. "My Dainty Cigarette" and "My Adored Cigarette," both from 1894, touched not too subltly on sexual themes. The latter began with: "As I lie on my sofa reclining, alone with my sweet little friend ..." and continued:

> The warmth of her ardor returning,
> Her form 'twixt my lips I can press,
> For me and me only she's burning
> Wrapped around in her little white dress.

By the early 1900s, the sexual theme had extended to include women in song. Anna Held, who sang "My Little Murad" in 1908 in New York, crooned:

> Whenever I'm blue
> There's one thing to do
> Make Love to my little Murad.

Victor Herbert's "Love is Like a Cigarette" (1908) was followed in 1913 by Sigmund Romberg's "Some Smoke (De La Fumee)." Later came "Smoke Rings," "The Great Cigarette Lady," and most familiar of all, Jerome Kern's "Smoke Gets in Your Eyes," from the 1935 film "Roberta."

In 1911 the job of "cigarette sampler" was established, whose duty was "to give away an occasional pack with tact and discretion." The industry recognized that a small investment could attract new lifetime customers. During the prewar period two all-American pastimes were linked: baseball and cigarettes. Cigarette trade cards at the games helped set up the first full generation of smokers.

Opposition was active

By 1910, a "nonsmokers' rights" movement of sorts started up. In April 1910, *Outlook* published an article by Twyman O. Abbott urging smokers to be more polite. Called "The Rights of the Non-Smoker," it said:

> There is a very large contingent in the community who do not use tobacco, and to whom it is not only obnoxious but detrimental to health. ... Formerly, and not so very long ago, it was the custom for a gentleman who wished to light his cigar in the presence of a woman to ask her permission. Nowadays, this act of courtesy is rarely extended.

Some lyrics of the day must have warmed the hearts of Lucy Gaston and her followers. Nasal minstrels sang out that, "Cigarettes and whiskey and wild, wild women ... will lead you to your grave." And surely Lucy must have endorsed the title of the 1912 song, "Cigarettes Must Surely Go."

By 1909, 12 states had enacted restrictive legislation and many towns and cities had done the same. A survey published in the *Chicago Tribune* showed that only Wyoming and Louisiana had failed to pass laws on the cigarette issue. The foes of cigarettes seemed to gain speed and enthusiasm in the early 1900s as the support for a national alcohol prohibition amendment fanned the fires of those who wanted to outlaw cigarettes next.

Some physicians spoke out, including Sir William Osler, Surgeon General Rupert Blue, Dr. Harvey W. Wiley (the driving force behind creation of the Food and Drug Administration) and a Dr. Tidswell who claimed that, "The most common cause of female sterility is the abuse of tobacco by males ... those countries which use the most tobacco have the largest number of stillbirths." As was the case in the 1890 campaign against cigarettes, the "educational" efforts were long on emotion and rhetoric, and short on facts. A typical poster in 1915 might read:

THE BOY WHO SMOKES CIGARETTES
NEED NOT BE ANXIOUS
ABOUT HIS FUTURE
HE HAS NONE
—David Starr Jordan

Connie Mack, Manager of the Philadelphia Athletics wrote in the *Scientific Temperance Journal* in 1913, "No boy or man can expect to succeed in this world to a high position and continue the use of cigarettes."

Education magazine told its readers in 1907, "There are in the United States today 500,000 boys and youths who are habitual cigarette smokers. Few of them can be educated beyond the eighth grade, and practically all of them are destined to remain physical and mental dwarfs." Often case histories were offered as the ultimate evidence in the antismoking groups' publications; for example: "Case 1: began habit at 4, taught by boys 6 and 7. Almost physical wreck now at age 13. Sight poor, voice like a ghost, hearing impaired. Steals. In first grade." Or: "Case No. 4: Began smoking at 10. Mind shattered at 14. . . . A worthless loafer now."

Concern about cigarettes focused on more than just children. During the pre-War period, many employers believed that cigarette smoking lowered productivity. Montgomery Ward, Sears Roebuck and hundreds of other large firms stopped hiring smokers.

Anti-smoking groups abounded. Members of the Non-Smokers Protective League of America, founded in 1911 by Dr. Charles G. Pease, a New York physician and dentist, regularly "arrested" smokers on subways and trains. In 1913, Miss Gaston and neurologist Dr. D.H. Kress opened a clinic in Chicago to cure smokers. The "cure" consisted of painting the palate with a silver nitrate solution and chewing some gentian root whenever the urge to smoke took over. Patients were to gargle with this substance after every meal, take warm baths and switch to a bland diet. The first cigarette after the mouthwash was supposed to make the smoker deathly ill. Another supposed remedy called No-to-bac, advertised for ridding all tobacco habits, was promoted with the motto: "Don't tobacco-spit your life away."

Lucy Gaston was recruiting prominent citizens to help her. Author Elbert Hubbard warned audiences that "Cigarette smokers are men whose futures lie behind them." David Starr Jordan, Chancellor of Leland Stanford Jr. University, warned that, "Boys who smoke cigarettes are like wormy apples that fall from the tree before they are ripe." Social leader Elinor Glyn, later a proponent of sexual freedom, noted that "Every smoke is a tiny drop of old age, so small that it is a long time that it goes unnoticed."

Despite strident campaigning against tobacco, cigarette sales increased dramatically. By 1911, when the great American Tobacco trust was ordered dissolved, sales neared 12 billion.

The doughboys take a drag

While the anti-smoking movement worked to attract support, the cause of cigarettes got a boost from the arrival of World War I. Tobacco companies quickly wrapped themselves in the flag to become a seemingly essential part of the war effort. Almost overnight, the moral taboo on cigarette smoking was overthrown. General John J. Pershing gave considerable prestige to tobacco when he cabled Washington, "Tobacco is as indispensable as the daily ration; we must have thousands of tons of it without delay." On another occasion the General stated vehemently, "You ask me what we need to win this war . . . I answer tobacco as much as bullets."

Army doctors sent home impressive reports of benefits for those wounded in battle. "Wonderful," one Army surgeon reported from France, "As soon as the lads take their first whiff they seem eased and relieved of their agony." In 1918, the War Department made the acceptance official: cigarettes became part of the daily ration. The War Industries Board estimated that Allied soldiers consumed 60 to 70 percent more tobacco during World War I than they did in civilian life. Since our government awarded contracts to cigarette manufacturers on the basis of pre-war domestic sales, Camels, which had 30 to 40 percent of the market, became the favorite among the boys in France. Day in and day out, the competition had to sit glumy by and watch thousands of new Camel smokers being created at government expense.

American doughboys came home smoking and singing the most famous soldier-song of the war:

> Pack up your troubles in your old kit bag and smile, smile, smile
> While you've a Lucifer to light your fag, smile boys, that's the style.

Lucy Gaston saw this coming and was among those who suggested that the "tobacco trust" had played on American patriotism to put dope in cigarettes so that servicemen would come home addicted for life. As a result, she felt she had to become even more of a political force and announced her availability for the Republican presidential nomination. She would follow the tradition of Lincoln by emancipating the nation from cigarette smoking.

Her platform was clean morals, clean food and fearless law enforcement. Her stated reasons for seeking the nomination were simple. First, she looked like Lincoln. Second, although men had made a good start in reforming America by abolishing slavery, she felt that it would take a woman to put the finishing touches on the moral uplift. Third, she had no husband to worry about. Fourth, she felt that the people of the country were solidly behind her in the "good morals movement" which was the paramount issue of the day. And Fifth, she noted proudly, she was a campaigner of 20 years' experience.

Six months after announcing her availability for the Presidency, she stepped aside "in favor of anyone who will endorse the moral reforms for which I stand." Apparently her choice here was William Jennings Bryan, but Warren G. Harding, a cigarette smoker, was elected.

Late in 1920 Ms. Gaston pleaded with President Harding to give up smoking, informing him by letter that "the United States has had no smoking President since McKinley, Roosevelt and Taft and Wilson all have clear records. Is not this a question of grave importance?" Harding died at age 59 of "stroke of apoplexy," very likely a cardiovascular complication of smoking.

In 1921, she organized a new National Anti-Cigarette League with the hope of creating a tide of support to follow passage of the 18th Amendment prohibiting the sale of alcoholic beverages in 1919. The war-demoralized anti-smokers had been encouraged by this amendment. "Prohibition is won; now for tobacco" declared evangelist Billy Sunday. The goal was clear: a 19th amendment to prohibit cigarettes.

But while legislative activity reached an all-time high in 1921 (that year 92 separate bills were debated by legislatures in 28 states), cigarette sales were skyrocketing. Tobacco companies were catering to the new demand from soldiers returning home, and more and more women were becoming interested in the habit.

The "ultra smart set" of women began to smoke around the turn of the century. By 1906, American "girl stenographers" were reported smoking cigarettes clandestinely. But it took the War, and the new independence of women as symbolized by the flappers of the 1920s, for women to begin smoking in public "with a trace of defiance," as one writer put it. By 1922 New York women were smoking openly on the streets and, as described in the next chapter, American advertising moguls saw and preyed upon this enormous market.

The anti-smoking group went downhill after 1921. As Werner put it in the *American Mercury*:

> At first—speaking in terms of decades—the public reaction to (the anti-smoking propaganda) was countrywide horror and revulsion. Then came, in turn, doubts and indifference; and lastly and most recently, discovery and disillusionment. The agitators had agitated, not wisely, but too well.

One state legislature after another repealed anti-cigarette statutes. By 1927 the anti-cigarette movement was both legally and functionally dead. Meanwhile U.S. production of cigarettes had increased by more than

5,000 percent since 1889. As Gordon Dillow wrote in *American Heritage* in the "Hundred-Year War Against the Cigarette," "Cigarettes benefited from that almost perverse quality of human nature that makes what is despised and outlawed by some people—particularly Sunday-school teachers and reformers—absolutely irresistible to others."

Anti-tobacco forces sought to penalize smokers calling for increasingly higher taxes on the commodity. Ironically, their ulitmate effect was exactly the opposite of what they intended. Few smokers were deterred by higher taxes, but governments appreciated and indeed became dependent on the enormous tax revenues from tobacco products.

The first federal excise tax on cigarettes had been imposed during the Civil War to raise money for that effort. The taxes were increased, then lowered after the War. Foes of cigarettes continued to press for higher taxes, and by 1919 the federal rate had jumped to 6 cents per pack. State taxes were first imposed in 1921. Thus smoking came to be a major component of the nation's budget.

On Camels and with Lucky Strikes

While Lucy Gaston was marching, some ingenious American businessmen were working to give America what it presumably wanted: a good-tasting, inexpensive cigarette. It was Richard Joshua ("Josh") Reynolds who provided this.

Reynolds had come to cigarette land with a great deal of experience in the use of tobacco. He had set up a tobacco plant in Winston, North Carolina, in 1875. In 1893 his company had promoted chewing tobacco as an additive-free, 100% natural product: "We continue our manufacture of tobacco . . . that has a natural sweet aroma, and cannot be improved on by the use of drugs. We use no flavors nor do we adulterate with cheap sweetening . . ." Ironically, this approach did not work very well, so Reynolds went in another direction, sweetening the product with saccharin.

One of Reynolds' first successes in the tobacco field was a pipe blend which he called Prince Albert. Overcoming his personal aversion to cigarettes (like James Buchanan Duke, founder of the American Tobacco Company, he never smoked them himself), Reynolds recreated a new brand which he called Red Kamel. The brand did not sell well, and he discarded it in 1909. But he didn't discard the name. Instead he modified it, and in 1913 introduced Camel cigarettes, complete with a wrapper featuring a picture of a camel, with palm trees, sand and two pyramids in the background. His Camels were an instant success. He didn't even have to offer coupons or premiums to sell them. (Indeed this was the end of coupons until Raleigh reintroduced them in the 1930s.) Smokers simply liked their taste and lightness. Reynolds, realizing that he had a winner, invested millions of dollars on advertising. One message was: "Tomorrow there will be more Camels in this town than in all Asia and Africa combined!"

Reynolds died in 1918, just a few years after his Camels encountered their first serious challenge. Camels had been the nation's Number 1 Smoke for more than a decade when Luckies moved in.

Luckies were the brainchild of George Washington Hill, son of the President of American Tobacco, Percival Hill. The younger Hill became intrigued with a newly developed product based on a blend previously used in plug and pipe tobacco under the name "Lucky Strikes." This had been particularly popular as plug during the Gold Rush. He had already made a name for himself, successfully promoting Pall Mall beginning in 1907. He introduced Luckies as a cigarette in 1916, complete with its now famous bull's-eye on the package.

Hill promoted Luckies by linking them with pleasant rewards—particularly the taste, smell and appearance of good food. "It's Toasted!" became the Lucky Strike slogan, and this explanation went with it: "The Burley tobacco is toasted; makes the taste delicious. You know how toasting improves the flavor of bread. And it's the same with tobacco exactly." Apparently the home-cooked image came to Hill when he saw in a processing plant that the heat used in making cigarettes was similar to that used in home cooking. Lucky's first advertising campaign shows a piece of toast with a fork stuck through it.

What did "toasted" mean? It didn't matter. It just sounded good. The fact that Luckies were "toasted" at first amused the competition who asked "Whose cigarettes weren't toasted?" But when Hill became more aggressive and asked, "You wouldn't Eat Raw Meat, Why Smoke Raw Tobacco?" the competition laughed less.

From the very beginning of his involvement in cigarette promotion and advertising, Hill proved his genius. His first step was the adoption of a fleet of automobiles, contraptions which then were a novelty. Each salesman was enabled to visit even the most remote areas to interest potential smokers and deliver stock to dealers. Hill worked out what may have been the most complete salesmen-routing system American business had yet seen. Innovation was his hallmark. Soon he adopted vacuum tins, cartons wrapped in glassine paper, dating of cases and cartons—and systematic inspections of dealers' stocks.

By 1910 Americans were smoking almost eight billion cigarettes, many with exotic foreign names like Fatimas, Meccas, Hassans, Helmars, Murads and Egyptian Deities. But as Luckies, Camels and a newly-promoted Chesterfield joined the scene, the use of foreign mixtures declined. "I'd walk a Mile for a Camel," bragged Reynolds. "They Satisfy," Liggett and Myers' Chesterfield assured us. "It's Toasted," Luckies continued to claim. The great cigarette brand war got into full swing. Even the trust-busting blow of 1911 (a time when there were widespread concerns about possible dangers of big business) did little to dampen sales. It might even be argued that trustbusting Teddy Roosevelt did the cigarette companies a favor by laying the groundwork for the competition of the 1920s and 1930s when he issued a dictum to the Supreme Court, commanding a complete and immediate dissolution of Duke's monopoly.

Cigarettes and health

Sales were booming, men and women were enjoying, and advertisers were getting geared up for the campaign of the century. There was certainly no reason to be concerned—even though for the first time in American history, a significant portion of men, and a few sly women, were regularly taking tobacco smoke into their lungs. Did anyone think seriously of lung cancer and other diseases then?

No, not really. Physicians for over a century had claimed that tobacco caused ailments such as colic, diarrhea, nausea, ulceration of the lungs, asthma, cough, heart pain, apoplexy (stroke), undernourishment, impotence and dulling of the brain. But all the scare talk boiled down to personal opinion or isolated observations.

There was even "good" medical news for the cigarette-makers during these pre-war decades. In 1906 in the *New York Times,* a doctor stated that modern women were drinking too much tea, and he advised that they take up smoking, as nicotine would counteract the stimulant in tea and prevent heart attacks. In 1918, the same newspaper carried headlines which read, "Surgeons Laud Cigarettes," and a text which advised that servicemen be given a full supply as "the effect of the cigarette is wonderful."

Pre-World War I crusaders had little evidence on which to base their claim that cigarettes were harmful. Tuberculosis and emphysema were the big threats of the day. And hazy statistics abounded. Crusader Charles Fillmore claimed that, "The New England Life Insurance Company found, after investigating records of 180,000 policy holders, that during a certain period, 57 out ot 100 nonusers of tobacco died; during the same period 95 out 100 cigarette smokers died." But no data indicated whether these smokers and nonsmokers were similar or different in other respects.

Smokers heard a great deal about "nicotinic amblyopia," which was described as "a peculiar weakening of the vision brought on by smoking." One Arthur D. Bush, M.D., reported in the *New York Journal* that smoking reduces mental efficiency by precisely 10.5 percent.

Reports of tobacco's dangers appeared and were quickly dismissed as unworthy of attention, which they usually were. One experiment claimed to have found that nicotine was harmful to dogs. But the popular press enjoyed dismissing this claim by writing, "An investigator connected with the John Hopkins University disposed of the nicotine-on-the-tongue of dog test by arguing that it would be interesting only if eating cigarettes was a general custom." Such data were labeled "moralist buncombe." In 1911, when a French physician was asked about some documented cases of cancer among smokers, he responded that the cancer would have occurred anyway, the tobacco may just have determined the location.

A review of the medical literature reveals only three significant events during the first two decades of the 20th century which did become part of the annals of valid smoking and health research. First, in 1912, Dr. I. Adler, in his book *Primary Malignant Growths of the Lung and Bronchi* wrote, "On one point, however, there is nearly a complete consensus of opinion and that is that primary malignant neoplasms of the lung are among the rarest forms of disease."

Second, Dr. Alton Ochsner who founded the Ochsner Clinic in New Orleans and who in later decades played a major role in alerting the public to the dangers of cigarette smoking, made mental note in 1910 during his junior year in medical school at Washington University in St. Louis, of a patient admitted to the Barnes Hospital with lung cancer. His teacher, Dr. George Dock, who was an eminent clinician and pathologist, asked the two senior classes to witness the autopsy because the condition was so rare he thought that they would never see another case as long as they lived. Dr. Ochsner was very much impressed by this event and became one of the first American physicians to blow the whistle when lung cancer became common.

Third, on May 28, 1916, the *New York Times* carried an unusually perceptive obituary. The death being reported was that of General Frederick Dent Grant, son of General Ulysses Grant. The headline read, "SAYS GEN. F.D. GRANT WAS CANCER VICTIM. DR. ABBE BLAMES TOBACCO HABIT, INHERITED FROM FATHER, FOR FATAL THROAT INFECTION." The facts regarding the death of the younger Grant were contained in an article by Dr. Robert Abbe, Senior Surgeon at St. Luke's Hospital who was with Grant when he died:

> One of our great national heroes smoked incessantly ... and suffered and died from the consequences of disease of his throat. His distinguished son, also a heroic figure in our army, adopted the same habit, smoked equally incessantly and suffered and died of the same terrible consequence. This is a heavy price to pay for the intemperate indulgence of such a throat-irritating and unnatural habit.

But even such a specific linking of premature death with the use of tobacco did nothing to halt the stampede of the tobacco companies and their customers. By 1920 the cigarette had a firm grip on a growing number of Americans, and for the next four decades, that grip tightened even more around their throats.

7

The Cigarette Hit Parade: 1920-1940

"The American people . . . are a bunch of saps that would smoke rat poison and bathe in chicken shit if you spent enough money advertising it."-T. P. Warham in the novel, American Gold, *by Ernest Seeman*

Cigarette marketing is a story of many ingenious businessmen, but two names stand out. First, there was George Washington Hill—sometimes called the "*enfant terrible* of advertising"—who took over the American Tobacco Company after his father died in 1925. Mr. Hill's commitment to Lucky Strikes was likened to a "missionary's devotion to Jesus." He had two special pets—dachshunds named Mr. Lucky and Mrs. Strike. He was rarely seen without his battered felt hat and a Lucky Strike between his lips. (Like many smokers, he was only 61 when he died of heart disease.)

Then there was Albert D. Lasker, President of the Lord and Thomas ad agency, often referred to as the father of modern American advertising. His clients included Kleenex, Pepsodent, Palmolive, RCA, Sunkist—and Lucky Strikes.

Considering the disastrous effects of cigarette smoking on health, it would be easy to condemn these men for hooking Americans on a deadly product. But Hill, Lasker and their associates, as well as almost everyone else, thought that cigarettes were safe and enjoyable. Furthermore, their activities embodied the spirit of free enterprise that dominated the rapidly industrializing United States.

Edward Bernays, an advertising man who worked with Hill, states in his memoirs:

> Had I known in 1928 what I know today, I would have refused Hill's offer. In the first place, cancer has been strongly linked with cigarette smoking. Furthermore, I no longer enjoy participating in or watching the kind of lethal competition that fascinated Hill and drove him to so many commercial excesses. With pitiless and ruthless force, he tried to dominate the market and destroy all competition.

Commenting on his role in persuading women to smoke, he noted it was "a beginning . . . I regret today." Later he became a senior member of the Board of Directors of Action on Smoking and Health (ASH), an antismoking advocacy group.

Emerson Foote, another executive who was a major promoter of Luckies during the 1930s, quit commercial advertising in 1964 to bring the message about cigarettes and health to the public. He had stopped smoking and became a member of a task force that made recommendations to the Surgeon General during the 1960s.

Lasker, too, demonstrated his great public spirit by establishing the Albert and Mary Lasker Foundation, an organization dedicated to supporting medical research.

Advertising's bag of tricks

By 1920, cigarette consumption was 665 cigarettes per capita, up from 310 in 1915—amazing numbers when one considers that at the turn of the century, cigarettes were still a novelty. Through the efforts of the ad men and the new social climate of "liberation" for women, that number soared to 1,976 per capita in 1940.

Cigarette ads in the '20s and '30s were clever, original, brazen, alluring, sometimes very funny—and extremely high-pitched. One would have had to be deaf and blind to avoid them. (An astute observer called the situation "advertising gone mad.") Women, youth—and other Americans who wanted to feel sophisticated, healthy, young at heart, carefree, and robust—became the targets of the advertisers. They cast their nets and pulled in millions of new smokers.

The great 20th century cigarette campaign began after World War I by capitalizing on the patriotic mood. A *New York Times* article set the stage by claiming that tobacco "affords true enjoyment; it helps our organism over many difficulties and over as many cares and hardships leading to a depressed state. It satisfies thirst and hunger, as we learned during the war."

Ads, of course, continued this theme. In 1918, all over the country, one could see drawings of the classic doughboy, leaning wearily against the side of a trench, clothes dirty, face streaked with soot, but with a knowing smile and a freshly-lit cigarette between his lips, and saying: "Murad— After the Battle, the Most Refreshing Smoke is Murad."

Many celebrities endorsed cigarette smoking for a fee (even though some were nonsmokers). Lucky Strike enlisted the noted opera singer, Madame Schumann-Heink, who for $1,000 became the first woman to provide public testimonial for the cigarette. Her endorsement was short-lived, however. After a number of colleges in the West cancelled appearances because of her participation in the ad, she withdrew her support for the cigarettes, and later actually denounced their use.

Many ads pictured not the endorser, but simply his or her hand—holding a cigarette, of course. As noted by *Printers' Ink* in August 1921:

> There is always something enormously interesting about hands, and especially about hands of well-known people. . . . Often, too, the hand tells something of the character of its possessor which may not be seen in any other feature or trait. . . . Nearly everyone will stop to examine the picture of the hand of a notable person.

"Lending a hand" to advertising were such prominent actors, producers and writers as Raymond Hitchcock, Leo Ditrichstein, Leon Errol, Thomas Meighan, Frank Bacon, Robert Mantell, Charlie Chaplin, William S. Hart, and Roscoe Arbuckle, Louis Joseph Vance and Thomas Mason, managing editor of *Life*.

Doctors, athletes and others even suggested that good health and good looks were the rewards of cigarette smoking. In 1927, American Tobacco mailed physicians a carton of 100 cigarettes and a questionnaire with a card carrying two questions, the first of which was "In your judgment, is the heat treatment, or toasting process, applied to tobacco previously aged and cured, likely to free the cigarette from the irritation to the throat?" The company then advertised that 18,000 American physicians had answered this question in the affirmative. Subsequent ads asserted that cigarette smokers "respect the opinions of 29,679 physicians who maintain that Luckies are less irritating to the throat than other cigarettes." To assure authenticity, the ad noted that the figures "have been checked and certified by LYBRAND, ROSS BROS. AND MONTGOMERY, ACCOUNTANTS AND AUDITORS."

One of the accusations that distressed tobacco promoters most was the charge that cigarettes irritated the throat. So an ad campaign was launched to counter this claim. In 1934, Philip Morris and Company financed an experiment on rabbits and concluded that the smoke from cigarettes which contained glycerine as a moistening agent caused the bunnies' irritation, while smoke from tobacco containing diethylene glycol "had only a slight and momentary action." Not surprisingly, Philip Morris cigarettes contained diethylene glycol. The company then hired 10 doctors to experiment with humans and advertised in the *Journal of the American Medical Association* that, "Patients with coughs were instructed to change to Philip Morris cigarettes. In three out of four cases, the coughs disappeared completely. This Philip Morris superiority is due to the improvement of diethylene glycol as a hygroscopic [moisture-retaining] agent."

Health claims grew even bolder. The makers of Camels claimed that they helped the digestive process. Advising people to "Get a Lift with Camel," later ads cited research at Yale which supposedly found that nicotine makes "the adrenal glands excrete adrenalin which makes the

liver and muscles pour their stored up sugar into the blood stream, where it becomes available for work, pleasure or refreshment."
Other ads were addressed to "hair mussers":

> If you catch yourself mussing your hair, biting your nails, chewing pencils—or suffering from any other of those countless little nervous habits—Get enough sleep and fresh air—find time for recreation. Make Camels your cigarettes. You can smoke as many Camels as you please, for Camels costlier tobaccos never jangle your nerves.

Variations on this theme included:

> "Are You a Key Juggler? Watch Out for Jangled Nerves . . ."
> "For years this has been no secret to those who keep fit and trim. . . . They know that Luckies steady their nerves and do not hurt their physical condition."
> "Not a Cough in a Carload."
> "They Satisfy."
> "Be Nonchalant."
> "I Smoke for Pleasure."

Tareyton ads even claimed that "They Steady Your Nerves." These ads usually showed "nerve strained" workers on the job, often with a Tareyton between the lips. Featured also were such stressed people as a newspaper reporter, an aviator, an officer of one of the big transatlantic liners, a deep sea diver and a novelist. Later one could observe under tension a train dispatcher, an engineer and a nurse.

Brand image

"Particular People" preferred Pall Malls. "Tobacco experts" preferred Luckies. Tired people got a lift from Camels, and irritable folks were advised to choose Philip Morris. (Do you suppose that an irritable, particular, tobacco expert who occasionally needed a lift needed to carry all four brands?)

Tobacco men developed new and ingenious ways to get their messages across. Philip Morris hired a small pageboy from the Hotel New Yorker and gained wide recognition. Little Johnny's "CALL FOR PHILIP MOrrrrrris" soon became the most familiar slogan on the radio waves. Philip Morris ads began to comment on the previous night's hockey games, "calling" the outstanding player. Reported one observer,

> It is known that these advertisements have quite an extensive following. At the games between periods a common topic of conversation among the fans is "Who do you think Philip Morris will call tonight?"

George W. Hill may be remembered best for his clever use of radio to promote Lucky Strikes. This included sponsorship of the Metropolitan Opera, radio programs such as "Information Please," "Hit Parade" (whose listeners could win 50 cigarettes by predicting the three most popular songs of the week), musician Eddy Duchin, and columnist Dorothy Thompson. Master Hill broke with tradition in other ways in an attempt to distinguish his cigarettes. He turned Pall Mall into a king-size cigarette (85 millimeters) compared to the regular one which was only 70 millimeters. His competitors followed his example soon afterward. As a *Fortune* editor pointed out, "People are talking, sales are responding, there's excitement in the air at 111 Fifth Avenue. For that feeling George Hill will pay almost any number of millions."

Critics respond

Cigarette promotion was a highly competitive science dedicated not only to hooking new smokers, but to persuading those already hooked to switch brands. Of course, all this hoopla over cigarettes did not get by without criticism. A 1929 *Commonweal* advertisement lamented the "conspicuous change in the advertising methods of some of the leading manufacturers of cigarettes..." and opined, though somewhat hopelessly:

> Somehow Europeans do not find it necessary to claim for a cigarette the most astonishing virtues. They do not suggest that it will improve the quality of one's soprano, relieve fatigue, or clarify the individual. They guarantee it no medicinal or dietary powers. They guarantee nothing except that it contains tobacco. We wish that some manufacturer in this country would try that—produce a good cigarette at a popular price and advertise it sensibly. He could even afford to disparage it a bit. He might say, "This is just a cigarette. It is neither a tonic nor a cough drop. If you smoke too much, it will result in a loss of weight and nervousness. For the sake of health, it is best not to smoke at all. But if you must smoke, and you want a cigarette, here it is."

A 1930 *Journal of the American Medical Association* expressed rage over the "modern tendency for advertisers of all kinds of merchandise to drag the health angle into their advertisements." Referring to American Tobacco's "survey" on Luckies and throat irritation, the Journal said, "The medal for the most horrible example would seem to go to the American Tobacco Company." Dr. James Tobey wrote in *Scribners Magazine*, "No more unreasonable and bigoted are some of the predatory, mercenary and rapacious commercial tobacco interests, whose sales methods and advertising ethics or lack of them, are, to put it mildly, definitely malodorous." Senator Reed Smoot (R-Utah) decried the "orgy of buncombe, quackery and downright falsehood and fraud."

But the ads continued, as they do today, and for the same reasons:

cigarettes were legal, and people wanted them. Unlike today, neither advertisers nor consumers were aware of the dangers involved.

The appeal to women

While a few daring women smoked cigarettes before World War I, "nice girls" didn't. Movies and plays highlighted the female villain by putting a cigarette in her hand. In 1922, an 18-year-old girl was expelled from Michigan State Normal College for smoking cigarettes. When she later brought a suit against the college president, claiming her individual freedom was violated, the Michigan Supreme Court upheld the expulsion!

The effort to persuade women to smoke was deliberately planned and executed in scientific fashion. Lorillard's ads for Helmar brand in 1919 were the first to show a woman with a cigarette in her hand (though she was never shown actually smoking). The ads featured oriental settings, with women lounging on divans and sofas, looking almost drugged, and oozing sensuality. This advertising approach was not considered effective and was soon dropped. In 1926, a lovely young thing in a Liggett and Myers ad didn't smoke but begged her companion to "Blow Some My Way." That same year, Bryn Mawr College lifted its 28-year-old ban, allowing its students (all female) to light up.

A year later, a Philip Morris ad for Marlboro cigarettes claimed that, "Women, when they smoke at all, quickly develop discriminating taste." Adding that their cigarettes were "as mild as May," the company implied that they were just right for the ladies. Curiously, Marlboro's current ads stress that the product is for the tough, macho, cowboy type.

The taboo on female smoking began to lift. In 1934, Mrs. Franklin D. Roosevelt was called the "first lady to smoke in public." A large part of increased cigarette use by women was linked with their growing freedom and independence. As noted by a 1932 survey:

> Among the visible results of these tendencies were the gradual concessions made by public opinion with regard to smoking by women, particularly in public places. While in Europe this change ... had taken place much earlier ... it was not before 1923 and 1924 especially that widespread smoking among women in this country began. ... It was about that time that the "flapper" had her day of bobbed hair, grotesquely dangling galoshes and skirts of extreme brevity. These younger non-conformists, encouraged by the national notice given their attire and manners, boldly began to pull their cigarettes in public.

The ads become increasingly aggressive and seductive. But it took the team of Hill and Lasker to finish the job, making cigarette smoking for women not only acceptable, but desirable. Working with Edward Bernays, Hill hired psychoanalyst A.A. Brill to predict what would induce women to begin smoking. Dr. Brill advised that, "Some women regard cigarettes as symbols of freedom." He then added: "Smoking is a sublimation of oral eroticism; holding a cigarette in the mouth excites the oral zone."

Armed with this information, Hill and Lasker hired a group of attractive models dressed in Lucky Strike Green to walk daringly up Fifth Avenue smoking cigarettes. Then the two promoters began looking for a new slogan for Luckies to replace the traditional "They're toasted." Hill told Lasker that he had met a woman who said she was 70, but looked 40. The reason she kept her looks, she said, was that she smoked cigarettes instead of eating candy. Hill thought of, "Reach for a Lucky Instead of a Bonbon." Lasker suggested a minor alteration and Lucky's new ad campaign was off: "Reach for a Lucky Instead of a Sweet."* The advertisement had everything. It appealed to women and to men on aesthetic grounds, it was a "conscience soother," and it even hinted that cigarettes were good for you.

Bernarr Macfadden's *Physical Culture* magazine and the anxiety of a nation too rich, too fat and worried about it was taking hold. Hill had the answer. The cigarette would be the companion of the figure-conscious American woman. Lucky Strike ads featured actress Helen Hayes claiming that Luckies accounted for the trim figure of the modern woman. Poster girl Rosalie Adele Nelson announced that she was a "lucky girl" to have found this pleasant way to all-around health.

A brief skirmish

Needless to say, candy manufacturers were not too pleased with this ad. Schrafft's outlawed the smoking of Lucky Strikes at their counter—and an all-out cigarette-candy war began. People who sold sugar were also unhappy. Cuba, Puerto Rico and the Philippines banned Luckies. Senator Reed Smoot had to deal with irate Utah sugar beet farmers. Legend has it that the Lucky ad almost bankrupted the candy business.

Candy companies sent out literature saying that candy was good, and hinting broadly that cigarettes were dangerous. Luckies responded by having Amelia Earhart state, "For a Slender Figure—Reach for a Lucky Instead of a Sweet," and Hill even sent to tobacco jobbers literature which claimed that "sugar is undermining the nation's health." Eventually Hill did modify the slogan to "Reach for a Lucky Instead." Then the Federal Trade Commission stepped in and officially ended the cigarette-candy war

---

*Fortune magazine carried a slightly different version of this story in December 1936: "One day ... Mr. Hill was driving home and saw, within a few blocks, a fat girl munching something and a svelte girl in a taxi lighting up a cigarette. He called Mr. Lasker whose copywriters reached into advertising prehistory and pulled out a Lydia Pinkham slogan of 1891, 'Reach for a Vegetable Instead of a Sweet' and Lucky Strike's most controversial campaign was born."

by prohibiting tobacco companies from selling cigarettes as a reducing aid, even by implication. Lorillard cleaned up by introducing a new slogan: "Eat a chocolate, light an Old Gold. And Enjoy Both. Two Fine and Healthful Treats."

Curiously, in the 1980s, cigarettes and candy joined forces to produce "Confectioner, Tobacconist, Newsagent," a trade publication emphasizing their common approach at point of sales. And the *United States Tobacco Journal,* subtitled the *Trade Journal of the Confectionary & Tobacco Industries,* recently editorialized that tobacco's fight is candy's fight, both the victims of "misguided do-gooders." This is most unfortunate. "Consumerist" attacks on sugar are not justified, but the attacks on tobacco most certainly are!

Media support

During the '20s and '30s, the media had a love affair with the cigarette. The *New York Times* of October 14, 1922, carried a front-page story concluding that the accuracy of work by a smoker could be adversely affected if he were deprived of his weed, but that while smoking might affect "fine reactions [coordination], there is no indication that the speed of complicated reactions is affected." A 1923 *New Republic* article assured:

> There is not the slightest foundation for the popular notion that the paper or the tobacco used in the manufacture of the cigarette contains any substance that is especially injurious to the human organism. Emphasis on the relative innocuousness of the cigarette is deemed justified by the persistence with which the misinformed strive to convey a contrary impression.

The *American Mercury* in 1925 featured a typical headline for its pro-cigarette article, "The Triumph of the Cigarette," reminiscing with some obvious relish, "Do you remember when they called it the coffin nail, and it was a common practice for austere gentlemen of Christian principles to snatch it from the fingers of young smokers. . . . What a change today."

In the March 1928 *Hygeia,* a publication which later became the AMA's *Today's Health,* C.S. Butler wrote a favorable article "On the Use of Tobacco in Prolonging Life." He said: "It is well to acquire a few bad habits in youth, so that as age advances one may have something to 'knock off' when the family physician, in his solicitude to prolong life, inquires into one's habits as to the use of alcohol, coffee, tobacco, food and profanity." The last line of Butler's article tells it all. Commenting on "My Lady Nicotine," he called it, "A goddess, at whose shrine the whole world worships, must have some good in her."

Assurances continued. Wingate Johnson in a 1932 *American Mercury* issue wrote, "There is little real evidence that smoking in moderation has any serious harmful effect upon the average individual."

Testimonials appeared frequently in *The New York Times*: "Smoked Seventy Years, Now Celebrating His Hundredth Birthday"; "Doctor Scoffs at Charges that Cigarettes Interfere with Health"; "Smoking Promotes Health, MDs say. It increases the flow of gastric juices and contributes to evenness of temper." In October 1926, a front page story reported the findings of a Johns Hopkins professor who had concluded that, "Smoking makes men more dependable because it acts as a sedative." True, this expert conceded, "Smoking does increase blood pressure slightly, but so does telling a good joke." His conclusion? Smoking was good for you. People believed him.

Smokers probably found a 1929 *Times* story entitled "Three-Year-Old Boy is Regular Smoker" somewhat amusing. This particular child, one Maurice St. Pierre from Waterbury, Connecticut, was said to have smoked two cigars, ten pipefuls and a pack of cigarettes each day. The story noted that he lighted his own, and that he had first taken up the habit when he was 1½ years old.

Where the print media left off, movies picked up. Humphrey Bogart and Lauren Bacall were the classic image—adult, suave, and in-the-know. Bogart without a cigarette? Impossible. It was always drooping from his mouth with Bacall whispering, "Got a match?" Bogart died of throat and esophageal cancer, a fact which Bacall, a smoker, neither mentioned nor related to his cigarette smoking in her book, *By Myself.*

The image of cigarettes was never perfect, however. In 1922 Carl Avery Werner wrote a "creed" for smokers, urging them to respect the rights of nonsmokers. The creed read as follows:

> Notwithstanding that those who derive happiness, comfort and good fellowship through the use of tobacco comprise 90 per cent of the male adult population of the United States, I fully realize that the majority, counting women and children, are nonsmokers and that among this majority there are some to whom the fumes of tobacco are not agreeable. I take pleasure, therefore, in observing the following rules of courtesy and consideration:
>
> 1. I shall not smoke or carry a lighted cigar or cigarette in any place or at any time where or when, either by placard or common understanding, smoking is prohibited.
>
> 2. I shall not smoke in any place or at any time where or when the fumes of tobacco are obviously annoying to others, even though such abstinence is not compulsory.
>
> 3. I shall not smoke in any passenger elevator, public or private.
>
> 4. I shall not smoke in a dense crowd of people, indoors or out, if I discover that my smoke is annoying some one near me who, owing to the circumstances, is unable to move away.

5. I shall not smoke in any home or any room wherein I am a guest without first making sure that smoking therein is agreeable to my host and others present.

6. I shall not smoke in the presence of any lady until I have been assured that she has no objections to my doing so.

7. I shall not approve of the use of tobacco by growing boys or girls.

8. I shall exercise caution in discarding the ends of cigars and cigarettes in order to preclude the possibility of fire.

9. I shall, in my enjoyment of the smoking privilege, be always considerate of those whose inclinations happen to differ from my own and always be guided by the finer instincts of true chivalry and American manhood.

10. I shall faithfully adhere to the foregoing self-imposed rules myself, and I shall urge others to do the same, that the days of tobacco may be long and its friends legion in the land of our fathers.

In November 1924, the *Reader's Digest* began what proved to be an ongoing educational campaign, first to convince readers to think carefully before they decided to smoke, and later, when the medical data began to come in, to convince smokers to quit. In an article entitled "Does Tobacco Injure the Human Body?" author Irving Fisher reviewed the opinions of a series of physicians on the effect of tobacco on health and concluded: "From every indication, it behooves the man who wishes to remain fit to omit tobacco from his daily schedule." This article contained information ranging from speculation and personal opinion to real scientific data. In February 1935, the *Digest* detailed some tips on how to give up the habit, including "eating sweets," "gradually cutting down one's rations" and "sheer will-power." In March 1936, in "A Burning Question," the *Digest* looked not at health, but what might be called "cigaretiquette," discussing how careless people were with the burning weed, and how much destruction it was causing.

A May 1929 issue of *New Republic* focused on the question of women smoking and concluded that women, unlike men, had very bad cigarette manners:

> When President Neilson of Smith College announced to his students the new rule restricting smoking to fireproof rooms, he closed this necessary but bound-to-be-unpopular address with this amiable comment: "The trouble is my dear ladies, you do not smoke like gentlemen."

There was a good deal of truth to what Neilson said. Smoking over the centuries was almost a ritual, guided by strict protocol. Men followed rules, including asking if they could light up. The *New Republic* article (by a female writer) complained:

> In some cities today it is next to impossible to purchase a chiffon evening

frock or a bit of lingerie except from the depths of a soft chair upholstered in green-glazed chintz, where the customer is surrounded by ashtrays and glowing cigarettes provided by a thoughtless management.

It concluded by describing a woman who was seen:

leaning on one elbow at a lace counter and puffing at a cigarette while purchasing yards of tulle frills, . . . a symbol of what the unchastened woman can do when she has not been taught to smoke like a gentleman.

The July 1938 *Reader's Digest* picked up on this theme:

Women haven't yet learned how to smoke, or when or where. . . . Look about in a restaurant: every woman in the place is sitting with her elbows on the table, one hand sticking up and awkwardly holding aloft a cigarette as if waiting for Buffalo Bill to shoot its end off. And who hasn't seen girls eating with a fork in one hand and a cigarette in the other? No man, not even a heavy smoker, would so ruin the taste of both food and tobacco. Women have brushed aside all traditions of courtesy and consideration regarding smoking. Men respect a few conventions, but who has ever heard a woman asking permission to smoke?

Also in 1938, in "Cigarette Holders Put to the Test," the *Digest* expressed concern about the fact that 162 billion cigarettes were smoked in 1937 even though "there is no physiological evidence that smoking does us any good." Citing nicotine as a poison ("drops of which can kill a dog"), the author recommended that if one must smoke, a cigarette holder would offer at least some protection.

Unhealthy rumblings

An impressive amount of speculation and some hard scientific evidence linking smoking with disease appeared between 1920 and 1940, but received little public attention.

In October 1920, University of Minnesota pathologist Dr. Moses Barron performed an autopsy on a 46-year-old male patient and determined that he had died of lung cancer. This seemed a bit odd, for another University of Minnesota pathologist had performed an autopsy two months before on a 42-year-old patient and found lung cancer. Still another death from the same cause was found later that month.

Dr. Barron had always thought that lung cancer was very rare. Sometimes a whole year went by without a single case among University of Minnesota autopsies. What was going on? Why were there three cases in a month? When additional cases came in, Dr. Barron decided to review the University's autopsy records to see if he could uncover a trend. He did, and reported it to the Minnesota State Medical Society meeting on August 25, 1921. Between 1899 and 1918, only four cases of lung cancer were identified at autopsy by University of Minnesota pathologists. There was one case in 1919. But during the single year from July 1, 1920 through June 30, 1921, eight lung cancer cases had turned up.

In 1922, Dr. John Harvey Kellogg published a book entitled *Tobaccoism: How Tobacco Kills* which pointed to smoking as the cause of lip, throat and mouth cancer, but the book received little attention, except among committed anti-smoking crusaders.

During the '20s and '30s, there was a sense of popular wisdom that cigarettes were not promoting health. As a 1921 issue of *Current Opinion* put it, "The weed has no standing whatever in the court of Science, Hygiene and Sound Sense." The objections to smoking in these decades involved shortness of breath, irritation, coughing, burning, nausea, hoarseness, and salivation. But nothing more serious than that was on people's minds. More important, there was no statistical evidence to back up these observations.

In a 1922 article in the British *Lancet,* Professor W.E. Dixon of Cambridge challenged physicians and scientists to find out more about the habit that was gripping the world:

> I venture to suggest that the collective sagacity of this Society and that of the medical profession as a whole could occupy itself with no subject more important to the nation than that of tobacco smoking.

In a July 30, 1927 letter to the editor of *Lancet*, Dr. Frank E. Tylecote wrote:

> As a clinician, I have remarks to make: (1) It might be assumed that the incidence of lung cancer is limited mainly to the working class. This is by no means the case; in Manchester we have lost several well-known public men from this disease in recent years. (2) I have no statistics with regard to tobacco, but I think that in almost every case I have seen and known of, the patient has been a regular smoker, generally of cigarettes.

The comments of Professor Pierre Schrumpf-Pierron of the University of Cairo are presented in the January 1929 issue of *Hygeia*: His research concluded that "excessive smoking is harmful and causes visible changes in the heart or the blood vessels." (Today we know that *all* smoking is excessive.) Later that year, *Current History* noted:

> The increase in cigarette smoking during the past decade and the vigorous advertising policies of the various tobacco companies have served to fix medical as well as lay attention upon the virtues and vices of tobacco.

An example of this growing interest can be found in a 1929 *American*

Review of Tuberculosis where Dr. Frederick L. Hoffman suggested that, in addition to cigarettes, the influenza epidemic of 1917-1918 might be a factor in the increase. At the 58th annual meeting of the American Public Health Association in 1929, enough concern was expressed about tobacco and health to warrant the passage of a resolution to include tobacco and tobacco products within the scope of the Food and Drug Act (something that never did happen).

In 1930 in *Lancet*, Dr. H.H. Sanguinetti noted the growing suspicion that tobacco smoking was a prominent risk factor in high blood pressure. That same year, a German researcher, Dr. Lickint, reported that of some 4,000 patients with bronchial cancer, 3,400 were men. He felt that the sex difference could be explained by smoking habits. He not only thought that cigarette smoking increased the odds of developing lung cancer, but he also thought that the product of burned tobacco might remain in the bladder and cause cancer there. (Some 35 years later, he was proven correct.)

In 1931, Dr. Hoffman had more to say about cigarettes and health—this time in stronger language:

> Possibly no phase of the highly complex cancer problem offers better opportunity for practical results than the general admitted correlations of excessive smoking habits to cancer of the buccal cavity, pharynx, larynx, and esophagus. Medical literature makes record of some outstanding illustrations, from Emperor Frederick II of Germany to General Grant, who are known to have died of cancer of the throat attributed to excessive habits of smoking.

Also in 1931, Dr. A.H. Roffo of the University of Buenos Aires isolated benzopyrene from tar formed by burning cigarettes and found that applying it to the tissues of experimental animals would cause cancer.

In 1932, in the *American Journal of Cancer* Dr. William McNally of the Department of Medicine at Rush Medical College, expressed grave concerns about the level of tar in cigarettes and its possible effects, and more important, linked cigarettes to the dramatic rise in lung cancer:

> Comparing the enormous consumption of cigarettes in 1925 to 1931 with the increase in pulmonary cancer, one is certainly led to believe that cigarette smoking is an important factor in the increase of cancer of the lungs.

In a desperate attempt to offer some protective advice, Dr. McNally recommended, "Cigarettes should not be smoked too short, as the last two centimeters retain most of the tar and other products of incomplete combustion."

In the same journal, Dr. Emil Bogen and Dr. Russel Loomis said that, "The clinical relationship between smoking and the presence of cancers of the lips, tongue, and buccal surfaces has been often noted. More recently, the increasing incidence of cancer of the lung has been blamed on tobacco." These writers recommended the use of a mechanical device to remove tobacco tar, a concept that became popular in the form of cigarette holders in the '30s and '40s.

In 1933, in the *Journal of the Medical Society of New Jersey*, Dr. W. Blair Stewart expressed his concern about the impact cigarette smoking was having on adolescents:

> If tobacco has a toxic effect on the heart, and in our mind there is no doubt that it has, is it not possible that some portion of our great increase of heart affections may be the result from the increased use of tobacco?

Despite these concerns, on November 25, 1933, *The Journal of the American Medical Association*, "after careful consideration of the extent to which cigarettes were used by physicians in practice," published its first advertisement for cigarettes (Chesterfield), a practice that continued for 20 years.

In 1935, Dr. Herman Sharlit noted, "It has been said frequently, and with some justification, that the medical profession broke faith with the public in failing to inveigh vigorously against smoking as a menace to health."

In a 1936 article in the *American Journal of Obstetrics and Gynecology*, Dr. Alexander Campbell expressed concern about the effects of smoking on the unborn child. He proceeded to survey obstetricians on this subject and found that the overwhelming majority agreed that smoking was dangerous for both mother and child during pregnancy.

In 1936 a German researcher reported that in a small study, 94 percent of the patients with cancer of the lung were heavy smokers. In a 1938 *Journal of the American Medical Association*, one of the first "hard data" studies on the increasing incidence of lung cancer was presented by Dr. Aaron Arkin and Dr. David Wagner. The doctors noted that primary carcinoma of the lung was one of the most frequent forms of malignancy in adults, with the right upper lobe being the most common site.

A 1938 edition of *Science News Letter* carried the headline, "Smoking Causes Cancer," citing the work of Drs. Alton Ochsner and Michael DeBakey (later a famous heart surgeon) of Tulane University School of Medicine. "Inhaled smoke, constantly repeated over a long period of time, undoubtedly is a source of irritation" to the lining of the bronchial tubes, the researchers reported. The newsletter then concluded that "10 to 15 out of every 100 primary cancers, not those that have spread from other cancers elsewhere, were lung cancer."

A few years later, Dr. Ochsner commented on the marked increase in patients with lung cancer and the fact that more than 95 percent of them smoked cigarettes:

> During my medical student days, I saw only one lung cancer case in four years. Today I operate on from two to five such cases every week. Now when I see a patient whose symptoms suggest lung cancer and who has been a heavy cigarette smoker, I make a tentative diagnosis of epidermoid lung cancer—or what has come to be known as Smoker's Cancer. Thus far I have been right in 98 percent of these diagnoses.

The doctors noted that lung cancer was usually a hopeless condition (as it often is today), with the only hope for cure being the removal of the entire lung and lymph nodes in the chest.

On March 4, 1938, a short but extremely significant article appeared in *Science*. In it, statistician Dr. Raymond Pearl presented the first tables based on family history data gathered by the Department of Biology at Johns Hopkins University which showed that: "Smoking is associated with a definite impairment of longevity." A few months later, a similar study by Dr. James Short and associates showed that "mortality markedly increased among heavy smokers." *Consumers Union* reported on the Pearl study in its July 1938 issue, but said that Pearl found that nonsmokers live "slightly" longer than moderate smokers, thus downplaying the findings.

In 1939, F.H. Muller, alarmed by the unprecedented increase in lung cancer in Germany, reported that of 86 patients with lung cancer, 83 of them smoked.

Although the dangers of cigarette smoking were becoming clearer to the medical profession, they received little or no press coverage in the public press. Why not? First, as Susan Wagner explained in *Cigarette Country*, large metropolitan newspapers were "fattening on tobacco advertising," and the last thing they wanted to print was bad news about one of their best clients. Second, at this point the majority of American men—including physicians and scientists who were coming in contact with these new and spectacular data—were smokers themselves, and human nature makes it very difficult to admit that something we do and enjoy is hazardous. And third, on the verge of World War II, other issues seemed more important.

The radio constantly sang the glories of the cigarette. The Prince of Wales even came up with a half-sized cigarette which he recommended for puffing between dances. The Great Depression had only a minimal effect on smoking because smokers would even give up food before cigarettes. In the pinch, they would buy "loosies"—one or two cigarettes at a time—giving merchants a little extra margin of revenue.

A 1939 *Fortune* magazine article reporting the first nationwide survey on the subject, found that 53 percent of adult men and 18 percent of adult women smoked cigarettes. These figures actually understated the marketing success of Madison Avenue. In the under-40 age group—for whom the advertising blast had been aimed, 66 percent of American men and 26 percent of women were cigarette smokers.

Smith College president William Allan Neilson seemed to convey the pre-World War II sentiment when he said of cigarette smoking, "It's a dirty, expensive, and unhygienic habit— to which I am devoted."

8
Luckies go to War

On the advantages of smoking: "If you smoke long enough, you will develop lung trouble, which will make you cough even when you sleep. Robbers hearing you cough will think you are awake and so will not try to steal your belongings."—Gene Tunney

During the '40s, the cigarette's grip on Americans tightened. Cigarettes were relatively inexpensive, as easy to get as a glass of water, and quite socially acceptable.

By 1940, brand names had proliferated, but Camel, Lucky Strike, Chesterfield, Philip Morris and Old Gold so dominated the market that their makers were charged by the United States government with conspiring to restrain trade. At trial the following year, the jury found American Tobacco Co., Liggett and Myers, and R.J. Reynolds guilty of: 1) combination and conspiracy to fix prices; 2) combination and conspiracy to create a monopoly; 3) attempting to achieve a monopoly and 4) achieving it. It was considered the government's most sweeping victory under the Sherman Anti-Trust Act. The companies were fined a total of $250,000. They appealed the verdict, but in 1946 the Supreme Court upheld it.

Despite the antitrust problem, tobacco companies prospered. World War I had boosted cigarette production from 18 billion in 1914 to 47 billion in 1918. World War II provided another giant boost.

Cigarettes seem to take on special status in times of combat. President Franklin Roosevelt declared tobacco an essential crop, and draft boards gave deferments to tobacco growers. As had Generals George Washington and John J. Pershing during previous wars, Douglas MacArthur asked that money raised for the war effort be used to obtain tobacco for his troops. Once, after a fundraising event, he reportedly said, "The entire amount should be used to buy American cigarettes which of all personal comforts, are the most difficult to obtain here."

During the war years, smokers worried that cigarettes might be rationed and that there might not be enough to go around. In 1943 a *Newsweek* story entitled "Cigarettes Scarce?" noted that, "Although in no instance yet reported has any United States community run completely out of cigarettes, in many places it is becoming increasingly difficult to get one or another brand." By 1944, when the prevailing price was about 15¢ per pack, cab drivers in big cities were selling Camels by the carton at twice this price. Reports from New York City grocery stores indicate that Manhattan's smokers were hoarding cartons of smokes. Cigarette production shot up over 300 billion per year, and consumption soared too. Domestic orders were not filled until the military demand was satisfied. In 1944 it was estimated that servicemen received 75% of the 300 billion cigarettes produced, tax-free of course.

By mid-decade, cigarettes were in such demand overseas—particularly in Germany—that they could be used as currency. According to a June 1947 issue of *Nation's Business*, when Secretary of State James Byrne took temporary leave from the Paris conference in 1946, the women in his party traded inexpensive cigarettes, obtained at military stores, for works of art and antiques at stores throughout Europe.

More sales pitches

Radio listeners during these years heard, again and again, the incomprehensible jabber of a tobacco auctioneer, followed by the clearly enunciated words, "SOLD AMERICAN!" Other advertising was intimately linked with the war effort, using models in military uniforms, surrounded by war paraphernalia. Chesterfield recommended that we "Keep 'em Smoking: Our Men Rate the Best!" Reynolds claimed that, "Camels are the Favorite! In the Army! . . . In the Navy . . . In the Marine Corps . . . In the Coast Guard!" Lucky's ads combined Morse code-like representations with sexual innuendo, saying that L.S./M.F.T. meant "Lucky Strike Means Fine Tobacco . . . So Round, So Firm, So Fully Packed, So Free and Easy on the Draw." Another ad, aimed at servicemen, featured Betty Grable, star of the movie, "Pin Up Girl," cooing that "With the boys . . . it's Chesterfield." (Miss Grable, a smoker, died of lung cancer in 1973 at the age of 56.)

But probably the most successful promotion involved the dramatic change of Lucky Strike packaging. Hill and his advisers never liked the green package. Ladies felt that it clashed with the colors of their dresses, and surveys showed that men didn't like the color either. But how could they gracefully change such a well-recognized package? By sending the green color to war, of course! In 1942 the green pack was replaced by a new white one heralded by the new slogan, "Lucky Strike Green Has Gone to War." The implication was that the pigments used to make Lucky Strike green were essential to the war effort. This was, of course, untrue. But it worked: sales increased by 38 percent in three months.

The basic sales approach of the '40s was relatively calm, involving what a tobacco executive called "dramatizing basic advantages." A leaflet for Lucky Strike Tobacco issued in 1896 had listed five basic reasons for the popularity of their product. It seemed that these themes were still successful:

Reason No. 1: "It is pure tobacco, containing the least possible quantity of sweetening and flavoring."

Reason No. 2: "It is wholesome, the nicotine being so eliminated that it may be used constantly, without risk of nervousness, indigestion and other troubles which frequently follow the excessive use of tobacco." Camel picked up on this during the '40s claiming "28 percent less nicotine in the smoke." Sano brand, among others began talking about "nicotineless cigarettes, pipes and tobacco," and Camels stressed, as they had in the previous decade, that their smoke "aids digestion" and improves your chances of "healthy nerves."

Reason No. 3: "It is a cool smoking tobacco and does not heat the pipe nor bite the tongue." Many of the ads of the '40s began talking about having more "coolness" and "less bite." Prince Albert stressed that it was "86 degrees cooler," and people were probably too busy lighting up to ask "cooler than what?"

Reason No. 4: "It possesses a fragrance that is enjoyed even by those who do not smoke, and always leaves a delightful odor in the room." In the '40s, Half and Half smoking tobacco, a product of American Tobacco Company, claimed, "Even your better half will enjoy the fragrance of Half and Half." Raleigh bragged of "sweet pipes."

Reason No. 5: "It retains its moisture and aroma in all seasons and in any climate." The cigar, cigarette and tobacco people came up with cellophane wrapping, and a "humidor pack" which became a major element in advertising. (Sobel reports that when soldiers in the Pacific theater complained that their regular cigarette packs became soggy in the damp climate, some ingenious manufacturers temporarily packed their product in converted Planter's Peanut cans to get around this problem.)

Not all ads during the '40s were subdued. The July 1942 *Reader's Digest* ("Cigarette Ad: Fact and Fiction" by Robert Littell), reported research laboratory findings on nicotine content, tars and resins and smoking time per cigarette. Old Gold was rated by the *Digest* as having the smallest nicotine and tar content of all brands tested, but the article stated that the tests showed the difference among the brands was negligible. Trumpeting this as a *Reader's Digest* endorsement of Old Gold, the company's ad men rushed to their typewriters, and Old Gold sales soared. "*Reader's Digest* exposes cigarette claims! Impartial tests find Old Gold lowest in nicotine, lowest in throat irritating tars and resins!" screamed the ads. But Old Gold's competitors had an answer. They noted that the brands "downgraded" in the *Digest* article were advertised in the *Journal of the American Medical Association*, which supposedly sanctioned only advertising claims which it considered authentic.

Brown and Williamson went back to stuffing premium coupons inside the wrappers of Raleighs. And industry leaders, one by one, entered the "king size" cigarette race; this larger cigarette was an instant success. The fact that it contained more tobacco was considered, of course, to be an advantage. The king size cigarette was part of a master plan to succeed in what *Business Week* called "A Drive for More Smokers"—and a related drive to get those people who smoked to smoke more.

The advertising claims did not go unnoticed by the Federal Trade Commission which filed sweeping complaints against Philip Morris and Company, Ltd., Inc., and R.J. Reynolds Tobacco Company. The FTC charged that Reynolds had falsely suggested that Camels were good for and aided digestion while Philip Morris had falsely represented and advertised that:

> Philip Morris cigarettes cause no throat or nose irritation; that when smokers changed to Philip Morris cigarettes every case of irritation of the nose and throat due to smoking cleared completely or definitely improved; that a smoker of Philip Morris cigarettes could depend upon and be assured of freedom from irritation of the mucosa due to something.

Smoke signs of the times

Cigarettes were everywhere in the 1940s. During that decade almost all ladies' pocketbooks came with neat little compartments for cigarettes.

The weed became a standard prop in movies and plays and, as Giles Playfair summed up in the April 1948 *Atlantic Monthly*, "Remarkable progress has been made in recent years in simplifying the job of acting." The cigarette conveyed the mood and the message. One could put together a virtual cigarette script, pointing out the cigarette's utility as an acting prop. One author suggested the following:

Caution and deliberation (particularly in the presence of an antagonist): Light a cigarette—or pipe or cigar—elaborately. Make as much as possible of extinguishing the match by holding it up before one's eyes and regarding the flame with interest before blowing it out.

Irritation: Flick the ash off a cigarette at frequent intervals, while tapping foot or drumming fingers.

Anxiety: Take quick and frequent puffs at cigarette, while moving briskly round stage or set. Discard a half-finished cigarette and straightaway light another.

Concentration (especially after moment of creative inspiration): Put aside a lighted pipe absent-mindedly.

Indecision (especially when in a tight corner): Take a long time to crush out a cigarette, using several superfluous motions in the process. Same effect can be achieved by knocking out a pipe.

Anger: Crush out a half-smoked cigarette—or cigar—impetuously. Then get up or, if already up, swing around.

Surprise: Have a cigarette—or more easily, in the case of a male character, a pipe or cigar—conveniently in mouth and remove with sudden, sharp gesture.

Acute distress or shock on receiving bad news: Crush out a half-smoked cigarette with awful finality. Stand quite still, keeping hand on butt of cigarette and head lowered, thus obviating need to reveal facial expression to audience.

Subtle threat of violence: Remove cigarette—or cigar—from corner of mouth with thumb and forefinger.

Self-confidence (especially after coming into money): Enter smoking cigar at jaunty, upward angle. If character is of humble origin, band should be left on cigar.

Disbelief: Exhale long puff of cigarette smoke slowly. If the disbelieved one is a shady character, blow smoke into his face.

Amusement: Exhale cigarette smoke with head tilted upwards, and give faint chuckle.

Shyness (especially man's shyness in presence of beautiful girl): Offer cigarette. Have difficulty in opening cigarette case. Have further difficulty in finding matches, and still further difficulty in lighting match.

Courage: Light a cigarette at every moment of danger. All female characters and characters of the Gentleman Crook school should take cigarette with their fingers from a cigarette case. Male characters of the Tough American Hero school, however, should take cigarette directly with mouth from a pack of cigarettes.

Fear (especially guilty fear): Try but fail—at least at first attempt—to light cigarette.

Passion in the raw: Put two cigarettes in mouth at same time. Light both. Then, with a possessive air, hand one of them to adored.

What would the movies of the day have been without cigarettes? Who could imagine Humphrey Bogart in his trench coat without a cigarette dangling from his lips while playing Sam Spade in *The Maltese Falcon* (1941)? "Spade put the cigarette in his mouth, set fire to it, laughed out smoke." And there was Lauren Bacall's opening line in *To Have or Have Not*, "Anybody got a match?"

Medical thunder in the distance

It is interesting to look back at the evidence on smoking and health during this period and wonder how it could have been ignored. In those days there was relatively little awareness of the relationship of health, lifestyle and the environment. The country was preparing for, fighting, and recovering from a World War. Spirits were high after the war—it was considered a time of good news, not bad. And the majority of adult Americans—including physicians—smoked.

By 1940, some 40 studies had been published on the health effects of cigarette smoking, mostly in foreign medical journals. During the '40s the pace of publication picked up in this country, and by the end of the decade, a few sophisticated and observant physicians began recommending that their patients, if they smoked at all, use some type of filtering device "just in case."

In the early '40s, many scientists found that the tars of tobacco could cause tumors in laboratory animals, thus confirming the earlier work of Sir Ernest Kennaway. Today such laboratory evidence involving a food additive or occupational chemical would cause a major protest and calls for the banning of the offending chemical. But in the '40s, such findings were considered obscure and unworthy of attention other than by a small group of research specialists. In addition to the further accumulation of evidence on smoking and health, during the '40s there was a type of backlash. A number of physicians—and many writers for lay audiences—soothed smokers with the most welcome of all words: "Don't worry..."

In 1940, in the *Journal of the American Medical Association*, Drs. John English, Frederick Willius and Joseph Berkson looked at the relationship of tobacco and coronary disease and concluded, "From the material comprising this study it appears that a greater incidence of coronary disease occurs among smokers than among nonsmokers." But there was considerable resistance to these conclusions.

In 1941, *JAMA* carried the research report of Drs. H.L. Friedell and L.M. Rosenthal who confirmed what had been known for years, namely that "Chewing tobacco is an etiologic factor in the development of cancer of the mouth." In June 1943, a chilling article in the *American Journal of Surgery* by Edwin Grace, M.D., emphasized what he called the "gravity of the smoking habit." To back up his charges, he referred to some of his own clinical experience:

> After having had an opportunity to observe, over a period of ten years, an unusually large series of patients with cancer of the lung, in two of the large municipal hospitals in New York City, two very distinct elements were noted in these patients; first, the patients were almost always men; second, they were heavy cigarette smokers and almost always inhalers. ... The gravity of this habit of smoking should be clinically emphasized.

Later that year, in the *Journal of the American Dental Association*, Dr. Grace directed his comments to dentists:

> Although positive proof at present is not available to establish a direct relationship between smoking and cancer of oral cavity, with the chemical isolation of a carcinogenic compound (benz(o)pyrene) from the tar of smoking tobacco, your profession must, I believe, realize the magnitude of its responsibilities.

There were some medical warnings for the military, too. Dr. John B.

McDonald was quoted in *Newsweek* in 1944 as saying that cigarettes may be dangerous to wounded soldiers: "The habit of giving an injured soldier a cigarette is not advisable if arterial injury has occurred." (The reason: In addition to the arterial spasm present in such cases, nicotine adds the aggravation of blood-vessel constriction, which may cause irreparable damage by decreasing blood supply to injured areas.)

A fascinating article on smoking and health appeared in 1947 in *Science Digest*. It was unique because it was written by Dr. Martin Gumpert, a physician who had pooh-poohed the idea of cigarettes being harmful until he suffered a heart attack as the result of a coronary occlusion. His opening comments probably typified the feelings of many smoking physicians of the day:

> Let me confess at the beginning that this article would have had an entirely different aspect a year ago when I smoked about two packages of cigarettes a day. Of course, I was aware of my indulgence and its possible harmful consequences. But I rationalized my addiction with the fact that I did not inhale the smoke and I often threw away a cigarette after a few puffs ... I must furthermore admit, with some embarrassment, that my transformation from heavy smoker to nonsmoker has profoundly influenced my scientific attitude toward tobacco.

Gumpert admitted that he had "more or less considered every nonsmoker a sort of faddist or crusader." But after his heart attack, he reanalyzed the literature on smoking and health, this time from the point of view of someone who was not committed to defending tobacco. His conclusions were similar to those of Harvard cardiologist Dr. Samuel A. Levine in his book *Clinical Heart Disease* (1945):

> When the question of smoking came up in former years, I used to tell the patients to smoke moderately, namely not more than eight cigarettes and two cigars daily. Now I am more inclined to urge omitting tobacco entirely.... We know that tobacco produces temporary depression of the T waves in electrocardiograms [indicating damage of the heart muscle].

But while these smoke signs of poor health were being launched in some medical journals, other journals carried data which would only console the worried smoker. A 1942 article in the *Canadian Medical Association Journal* concluded:

> Perhaps we are justified in saying that smoking is popular because it is pleasant, soothing, contemplative, and companionable. Like alcohol, it is a means of escape, without alcohol's dire potentialities for disease ... There is an old adage—*In Vino Veritas* ["In wine there is truth."] May we suggest a new one. *In Fumo Caritas*. ["In smoke there is charity."]

A 1947 *JAMA* article by Dr. Robert Levy and his colleagues concluded that:

Except in susceptible persons, smoking cigarettes causes only slight changes in the circulation and does not increase significantly the work of the heart. Because of the enjoyment afforded and the emotional satisfaction obtained, patients with inactive forms of heart disease may be permitted to smoke in moderation.

Echoing these conclusions, a 1948 *JAMA* editorial concluded:

From a psychologic point of view, in all probability more can be said in behalf of smoking as a form of escape from tension than against it. Several scientific works have been published that have assembled the evidence for and against smoking, and there does not seem to be any preponderance of evidence that would indicate the abolition of the use of tobacco as a substance contrary to the public health.

With this philosophy, *JAMA* rationalized its continued publication of cigarette advertisements, noting that complaints about these ads from physician readers did not exceed a dozen annually.

In between these two viewpoints on smoking and health were those who hedged their bets. In 1944 Dr. Clarence Cook Little, then Managing Director of the American Cancer Society, stated:

Although no definite evidence exists concerning the relationship between the use of tobacco and the incidence of lung cancer, it would seem unwise to fill the lungs repeatedly with a suspension of fine particles of tobacco products of which smoke consists. It is difficult to see how particles can be prevented from becoming lodged in the lungs, and when so located how can they avoid producing a certain amount of irritation?

During the 1950s Dr. Little became director of the Tobacco Institute's Research Council and a staunch defender of cigarettes!

Assurances that nothing was "certain" about cigarettes causing disease were picked up in the popular press. In "The Truth About Tobacco," published in the *American Mercury* in 1943, Robert Feldt, M.D., said:

If you are in good health, and use tobacco moderately, you needn't worry much about your smoking . . . It is easy for reformers to dismiss the tobacco problem by saying "smoking never did anyone any good" but the satisfaction that millions of confirmed smokers derive from a cigarette, pipe or cigar, must not be overlooked.

Science Digest in 1941 answered the question "Does Smoking Injure Health?" by saying, "In general physicians hold that smoking produces no apparent injury in sound individuals. . . . Undoubtedly the pleasure so derived over-balances the harm."

Standing almost alone in this early crusade against tobacco were the editors of the *Reader's Digest*. Although still handicapped somewhat by lack of hard data, they presented what they knew and offered moral boosts to those who wanted to give quitting a try. In 1941 they published "Nicotine Knockout, or the Slow Count" by Gene Tunney, former heavyweight boxing champion who was at that time a Lieutenant Commander in charge of Navy physical training and athletics. He advised that cigarettes contained nicotine and other toxic substances, and that no people who wanted to be healthy should use them. He ended his piece by offering three good reasons for smoking:

> First, if you smoke enough tobacco, you smell so strong the dogs will never bite you. Second, if you smoke long enough, you will develop lung trouble, which will make you cough even when you sleep. Robbers hearing you cough will think you are awake and so will not try to steal your belongings. Third, if you smoke as much as you can, you will have many diseases, and will die young.

Later in this decade the *Digest* challenged some of the claims made in tobacco ads, reminding readers of the foolishness of smoking and giving courage to those who would want to try to live without cigarettes. Asking in an August 1944 article "Are you a Man or a Smokestack," J.P. McEvoy detailed his fight for independence from cigarettes, his eventual victory and "the moral glow from conquering an enslaving habit [all of which] add up to the most exhilarating satisfaction in the world."

As 1950 approached, those who had contact with the world of smoking and health knew it was deeply enmeshed in conflict and controversy. A few hundred medical studies had revealed bad news about tobacco. But these studies were small and not well controlled by today's standards. Often they were brought to public attention by individuals who were either pervaded with a moralistic anti-cigarette bias or were committed to defending any attack against the cigarette industry. However, the time had arrived for a systematic and organized approach to research.

In 1949 the American Cancer Society commissioned the first of many studies by awarding a grant to Dr. Evarts A. Graham, the first surgeon to cure a case of human lung cancer by removing the affected lung, and one of his medical students, Ernst L. Wynder of the Washington University School of Medicine in St. Louis. Dr. Graham (a cigarette smoker himself) was not enthusiastic about the hypothesis that the increased incidence of lung cancer was related to cigarette smoking, and even noted that one could draw a similar correlation between increased lung cancer and the sale of silk stockings. But the research began—and led to a medical avalanche that shook the world.

9

The Evidence Mounts

"For thy sake, tobacco, I would do anything but die."—Charles Lamb, in A Farewell to Tobacco, *1830*

In 1775, in his book, *Chirurgical Observations,* London physician Percival Pott noted an unusually high incidence of scrotal cancer among chimney sweeps. He suggested a possible cause: beginning at a very early age, the sweeps climbed up narrow chimneys. Their hygiene was very poor, so that soot accumulated on their scrotums, eventually leading to "a painful and fatal disease." Pott recognized that an environmental cause of cancer was involved. Two centuries later, benzo(a)pyrene, a powerful carcinogen in coal tar, was identified as the culprit.

In 1849, John Snow reported a sudden increase in the number of cases of cholera in London. After some detective work, he concluded that the victims were not evenly distributed around the city but lived in one geographic area. Having made that observation, he did not take long to implicate the sewage-contaminated water flowing from the infamous Broad Street Pump. The causative agent was found years later to be bacterial.

In 1966 Boston gynecologist Howard Ulfelder saw a 16-year-old girl with adenocarcinoma of the vagina, a rare disease usually occurring in women over age 50. During the next few years, he and his colleagues encountered seven more girls aged 15 to 22 with the same disease. All of their mothers had taken stilbestrol during the early months of pregnancy. Thus was discovered another environmental cause of cancer.

In 1973 Dr. John L. Creech, a Louisville surgeon who helped care for B.F. Goodrich tire workers, casually mentioned to the plant physician, Dr. Maurice N. Johnson, that he had recently seen two cases of angiocarcinoma of the liver. This disease is so rare that it was unusual to have more than 25 cases reported per year in the entire country. A few weeks after this conversation, a third case of angiocarcinoma was identified in Goodrich workers. These two astute physicians correctly concluded that vinyl chloride—to which the workers had been exposed—was the causative agent.

The above cases illustrate epidemiology, the science of the cause and distribution of disease in human populations. In each case, alert observations led to identification of the cause.

Cancer and the concept of "cause"

With communicable disease, only a few days or weeks usually elapse between the victim's exposure and the development of the disease. Not so with cancer, which can take 10, 20, 30 or even more years to develop. This makes the study of cancer causation more complex than the study of infectious diseases. When Dr. Snow went to the scene to investigate the cholera epidemic, most of the people who had been there the week before were still around; and so was the pump. But when a cancer epidemiologist begins searching for leads, the equivalent of the pump may be gone, and the victims dispersed or perhaps dead from other causes.

Not everyone who drank from the Broad Street pump in 1849 developed cholera; nor were all existing cholera cases explained by exposure to water from the pump. But it was obvious that something in the pump water caused people to become ill. No elaborate statistical exercises were needed to prove the relationship between contamination and illness. However, *cancer's long latency period necessitates that the concept of cause be based on statistical association*—that in the absence of exposure to the factor, the disease in question would have occurred less frequently.

Tobacco apologists claim that the medical evidence against tobacco is "all statistical" and that the word "cause" cannot be applied until an exact biological mechanism is fully identified. From the public health viewpoint, this is ridiculous. Dr. Snow didn't know how the water of the Broad Street pump caused disease, but he did know how to protect people's health by shutting off the pump. Nor did William Jenner know the cause of smallpox in 1796 when he recommended vaccination with cowpox. He only knew that milkmaids who had previously had cowpox were immune to smallpox. This was a purely statistical association. The smallpox virus was not discovered until the early 1900s—over a century after the disease had been brought under control in the developed world.

The indictment

To establish environmental causes of human disease, epidemiologists apply rigorous standards. The process is much like putting together a puzzle. Before cause can be determined, the pieces must fit together. Above all, there should be consistency of evidence—that is, not too many puzzle parts that don't fit. Epidemiologists ask questions like:

1. Is there a biological hypothesis that might explain why a factor causes cancer? In the case of cigarettes, the question might be whether there is a biological reason *why* inhalation of cigarette smoke into the lungs might cause lung cancer.
2. Have individuals with a given disease had greater exposure to the factor in question than individuals who do not have the disease? In other words, are victims of lung cancer more likely to be smokers than are people without lung cancer? This approach (*retrospective* investigation) starts by identifying individuals with the disease in question and then trying to determine how they differ from individuals without the disease.
3. Do individuals exposed to the factor in question have a higher incidence of the disease? In other words, over a period of years, will more in a group of smokers develop lung cancer than in a comparable group of nonsmokers? This is the *prospective* epidemiological technique.
4. Are people exposed to greater amounts of the suspected factor more likely to get the disease than those exposed to lesser amounts of it? In other words, is there a dose-response relationship? Are heavy smokers more likely to develop lung cancer than light smokers?
5. Do populations not exposed to the factor develop the disease less frequently? Do Mormons and Seventh-day Adventists, whose religion prohibits cigarette smoking, have lower rates of lung cancer?
6. What happens when exposure to the suspected factor is discontinued? Do people who stop smoking diminish their odds of developing cancer?
7. Does this conclusion of causation make sense in terms of time trends? For instance, is the hypothesis that cigarettes cause cancer consistent with other facts we know about the incidence of cancer?
8. Are the conclusions from human studies consistent with those of animal studies? In this case, do the components of tobacco cause cancer in laboratory studies?

Cigarette smoking as a cause of cancer has probably been studied more intensely than any other subject in the history of epidemiology. *By 1953 all of the above epidemiological criteria had been satisfied.*

A unique dilemma

Despite the rumblings of danger, Americans entered the 1950s, cigarette in mouth, full of enthusiasm and confidence about their habit. By today's standards there was already enough evidence for the government to sound a public alarm and take regulatory action to protect Americans from what was clearly a hazardous product. But the cigarette-health issue was a unique one.

First, for smokers aged 40 or younger, cigarettes were as American as apple pie. Throughout their lifetimes, they had seen people smoking, had encountered innumerable ads praising the habit, and had watched movies in which cigarettes were used to communicate glamour, sophistication and confidence.

Second, cigarettes offered pleasure and relaxation to millions of Americans who liked them and became addicted to them. They were not easy to give up. An oft-quoted bit of public health wisdom states that if spinach had been the guilty leaf, there would quickly have been one less vegetable in our markets. But millions of Americans considered cigarettes indispensable.

Third, tobacco was an important economic commodity. By 1950 the cigarette industry had a firm grip on the country, with millions of Americans directly or indirectly dependent upon it for income. Unlike spinach, cigarettes could not be eliminated without creating economic shock waves.

Fourth, while there had been a substantial number of medical studies prior to 1950, none had followed the rigid scientific protocol needed to establish causation. The early literature on smoking and health was largely preliminary in nature. Researchers had become alarmed about the rising incidence of lung cancer. Studies on small groups of people had noted a link with cigarettes, and other studies had shown that tobacco tar could induce cancer in animals. However, a more comprehensive approach was necessary for the indictment to become a conviction.

Retrospective studies

The results of the first large-scale research on smoking published in an American medical journal appeared in the May 27, 1950 *Journal of the American Medical Association.* In "Tobacco Smoking as a Possible Etiologic Factor in Bronchiogenic Carcinoma," medical student Ernst Wynder and Dr. Evarts A. Graham described their study of whether hospitalized men with lung cancer were more likely than a control group to have been cigarette smokers. Identifying 605 men with bronchiogenic cancer, Wynder and Graham found that 96.5 percent were smokers while only 73.7 percent of men without cancer were smokers. They concluded that, "Excessive and prolonged use of tobacco, especially cigarettes, seems to be an important factor in the induction of bronchiogenic carcinoma." Dr. Graham himself began a concerted effort to stop smoking.

In September 1950, the *British Medical Journal* carried a preliminary report on smoking and lung cancer by Drs. Richard Doll and A. Bradford Hill. Between 1922 and 1947 the annual number of deaths from lung cancer had increased roughly 15-fold—the most striking increase ever recorded by the Register General in England. Doll and Hill asked 20 hospitals in London to notify them of all patients admitted with lung cancer. Examining the smoking rates for patients with and without lung cancer, they obtained data startlingly similar to those of Wynder and Graham. During the next few years, more than a dozen other investigations yielded similar results.

Prospective studies

The prospective approach is more costly and takes longer to carry out than the retrospective approach, but yields evidence that is more specific and valuable. In 1952, Drs. E. Cuyler Hammond and Daniel Horn of the American Cancer Society began a massive prospective study to observe what happens to comparable groups of smokers and nonsmokers. Using 22,000 trained volunteers, they enrolled over 187,000 men between the ages of 50 and 69 and used detailed questionnaires to determine their health status and smoking habits. The researchers anticipated that three years would be needed before significant differences between smokers and nonsmokers would be detectable. But after 22 months, they decided to take a preliminary look.

Dr. Hammond was a 4-pack-a-day man. Dr. Horn smoked one pack a day. As the IBM cards snapped out of the sorter, both of them became so alarmed that they switched to pipes. It was obvious that the overall death rate of smokers was 1½ times that of nonsmokers. The death rate from cancer for men who smoked a pack a day or more was 2½ times as great as for nonsmokers; smokers showed 5 times the death rate from lung cancer alone and twice the death rate from heart disease. Even men who smoked less than half a pack a day had significantly higher death rates than nonsmokers—and so did cigar and pipe smokers.

Also in 1954, smokers and the tobacco industry got a second big dose of bad news. Drs. Doll and Hill had gathered information on smoking behavior from 40,000 physicians aged 35 and older. After the questionnaires were in, the researchers kept track of the doctors for 4½ years, obtaining death certificates whenever deaths occurred. The new study concluded that, "Mild smokers are 7 times as likely to die of lung cancer as nonsmokers, moderate smokers are 12 times as likely to die of lung cancer as nonsmokers, immoderate smokers are 24 times as likely to die of lung cancer than nonsmokers."

More pieces of the puzzle

In 1953, further research by Wynder and Graham revealed that cigarette smoke condensate ("tar") could cause cancer in mice. Relating this finding to the mounting evidence in humans, Dr. Alton Ochsner predicted that, "In 1970 cancer of the lung will represent 18 percent of all cancer . . . one out of every 10 or 12 men." Ochsner's prediction was amazingly accurate. In 1970, lung cancer accounted for 19.7 percent of cancers.

In 1957 in the *American Journal of Obstetrics and Gynecology*, Dr. Winea J. Simpson asked what effects smoking might have on the unborn child. The incidence of premature births and of all the complications that go with prematurity was twice as great for smoking mothers as it was for nonsmoking mothers. Simpson's paper confirmed that children of smokers are not only born early, but also weigh less and are more likely to be stillborn or die within one month of birth. Later it was recognized that maternal smoking during pregnancy increases the chance of miscarriage.

The evidence continued to mount. What once was the "lung cancer scare" now implicated heart disease and a whole range of other serious diseases. Indeed, the list of cigarette ills became so extensive that tobacco industry apologists tried to arouse public skepticism by saying, "No one substance could cause that many diseases."

In 1956, at the request of the U.S. Surgeon General, a scientific study group was set up to review the evidence. Organized by the National Cancer Institute, the National Heart Institute, the American Cancer Society and the American Heart Association, the group evaluated 16 different studies and concluded that *there was a relationship between smoking and lung cancer.* In March 1957 the group's official conclusion was released: "The evidence of cause-effect relationship is adequate for considering the initiation of public health measures."

That same year, the British Medical Research Council reached the same conclusion; and Dr. Wynder and colleagues began studying what happened to Seventh-day Adventists, a population group who did not smoke because of religious reasons. Dr. Wynder's study analyzed all patients diagnosed with cancer and coronary heart disease during the previous five years in eight Adventist hospitals, five in California and one each in Illinois, Massachusetts and Washington, D.C.

The proportion of Adventists in the total patient population was about 1 out of 12. Thus, if all other things were equal, $1/12$th of the cases of cancer and heart disease should occur in Adventists. This ratio held for cancers of the colon, rectum and prostate, but not for lung cancer. Instead of the "expected" 10 or 11 cases, there was only one—and that patient, age 63, had smoked for 25 years before joining the church at age 50.

Cigarettes and the popular press

Newspapers and newsweeklies were generally dutiful in reporting the results of each major study as it appeared. But almost all of the articles "balanced" by citing a tobacco industry spokesperson or by using the qualifier "excessive." Saying that "excessive" smoking is harmful makes a statement seem less threatening. Unfortunately, the medical definition of "excessive" eventually turned out to be just about the same level of consumption of cigarettes most Americans were smoking at that time.

John Kenneth Galbraith, Professor of Economics at Harvard sent an interesting letter to *The New York Times* about its "balancing" of a news article. The article had reported findings of Dr. Harold Dorn, who had followed 200,000 veterans and found that the death rate for heavy smokers was twice that for nonsmokers. Dr. Galbraith wrote:

> Your news story . . . carried several paragraphs of a statement by Timothy V. Hartnett, head of something called the Tobacco Industry Research Committee which said it wasn't so. . . . While you give considerably more space to Dr. Dorn than to Mr. Hartnett, you treat the statements of both with equal respect. Does not this seeming impartiality mean, in fact, that you are allowing Mr. Hartnett to use you for his own purposes in a rather outrageous way? Shouldn't you make it wholly clear that you are not equating the work of a careful researcher extending over years with the press release of an industry spokesman?

A 1959 article by Davis Cort in *The Nation* was more critical of the newsweeklies:

> *Newsweek's* handling of the cigarette-cancer connection is the familiar one of confusion by verbosity. *Time's* story, much more thoroughly researched, nevertheless repeats this technique, burying the first mention of cigarettes under 650 lines.

Very few magazines gave consistently high priority to the topic of tobacco and health. A few at the other extreme pooh-poohed the data, but most took a middle-of-the-road position. Standing out from the crowd was the *Reader's Digest,* which had been questioning the advisability of cigarette smoking for many years. In 1950, the *Digest* published "How Harmful Are Cigarettes?" by Roger William Riis, warning that cigarettes might cause heart disease as well as cancer. Although acknowledging that the case was not completely proven, he noted:

> When I began research for this article I was smoking 40 cigarettes a day. As I got into the subject, I found that number dropping. As I finished the article I am smoking ten a day. I'd like to smoke more, but my investigation of the subject has convinced me that smoking is dangerous and worse—stupid.

At the other end of the spectrum was *Coronet,* which in 1950 published "The Facts About Cigarettes and Your Health," by Henry W. Mattison and John Schneider. This article began by comparing the bad news about cigarettes to "such scares as Orson Welles' famed broadcast of an Invasion from Mars . . ." The authors noted that, "Never before, in fact, have the prophets of doom so diligently exposed the alleged evil effects of tobacco." The message: Keep on Smoking, America. In 1959, *Coronet* published a tender and mellow piece on the history of smoking without once mentioning the health consequences of smoking cigarettes.

The *Digest* kept close tabs on the medical literature, offering tips on quitting or smoking in a "safer" fashion, featuring testimonials from people who did kick the habit, and issuing warnings on how careless cigarette smoking could lead to fires. ("So You Want to Burn to Death?" was published in September 1959.) But probably the most important article the *Digest* ever did on cigarette smoking was one by Roy Norr entitled "Cancer by the Carton" (December 1952). In a frighteningly convincing style, the author summarized the data linking cigarettes and lung cancer. The article was only two pages long—but extremely powerful. Millions of American smokers could avoid the issue no more.

An accurate and helpful article by Charles S. Cameron, who was then the medical and scientific director of the American Cancer Society appeared in the December 1957 *Ladies' Home Journal*. This article is remarkable because it discusses the data implicating cigarettes and disease much more fully than has any subsequent issue of the magazine.

U.S. News and World Report showed interest in the cigarette health issue in a balanced, informed way. In 1950 it gave extensive coverage to a report from the Federal Trade Commission which noted that "Cigarette smoking is not good for the individual. All cigarettes contain harmful substances. No brand is any better in this respect than any other brand." In 1954 the magazine carried two detailed interviews with Dr. E. Cuyler Hammond, Director of Statistical Research at the American Cancer Society.

New Republic was similarly outspoken, showing open hostility toward the cigarette industry's efforts to dismiss medical data. In 1957, *New Republic* recommended that government and private agencies launch a publicity campaign on the hazards of smoking.

Nation deserves mention, too. Accompanying a 1953 article by Dr. Alton Ochsner, the editors stated that, "*Nation* has never crusaded against smoking. Its position was that following the first significant studies showing a relation between cigarettes and lung cancer, cigarette smoking had to be regarded as a public health problem." In 1962, the magazine carried a cover story by Abraham Lilienfeld called "The Case Against the Cigarette."

Consumer Reports seemed surprisingly hesitant to take sides. Its editors today are particularly conservative—recommending that readers avoid taking risks, even hypothetical ones. But through the '50s, they hedged somewhat on smoking. In February 1953, for example, an article concluded with words that could have been interpreted as an endorsement:

> Smoking is an activity that serves to reduce the inner nervous tensions and strains resulting from man's exposure to the stresses and responsibilities imposed by society. It helps him to perform more effectively in his work and personal relationships. . . . As for cancer of the lung, while it has not been conclusively proved that heavy smoking is a major factor, the evidence for such an indictment is very suggestive.

In 1954, *Consumer Reports* discussed some doubts about the cigarette-cancer link, but did conclude, "The evidence does appear to be so strong that consumers would be well advised to decide whether to start smoking or to continue smoking." But in February 1955, they hedged some more, quoting a prominent physician who noted that the "cigarette theory is almost entirely based on statistical data having at best circumstantial value and being in part of questionable origin." The article concluded that cancer of the lung may be "partly" due to "excessive" cigarette smoking. As late as March 1957, another article concluded:

> The cancer linkage is still not clear . . . for anyone to argue that everyone should stop smoking because of its hazards would be highly unrealistic. . . . The stimulating or comforting effects of tobacco may be so valuable to some persons that they are willing to risk whatever physical harm may be associated with the habit.

Tobacco industry reaction

Cigarette executives knew of the growing medical data indicting cigarettes as soon as it began to appear in the '20s and '30s. They addressed some minor problems—like the charge that cigarettes irritate the throat and lead to coughing—but ignored the larger issues. In the early '50s, there was hardly a peep from the industry. As noted in *New Republic*, "Even the old families have been shaken. Philip Morris has retreated from gloomy reality to find solace in its new snap-open pack."

But in December 1953, an article in *Business Week* noted that "fast-paced events loosened up for the first time official tongues of the tobacco industry, which up until now has preserved a rigid silence on lung cancer." Paul M. Hahn, President of the American Tobacco Company (Lucky Strike, Pall Mall and Tareyton) issued "reassurance to the public," scoring the "much loose talk subject." E.A. Darr, president of R.J. Reynolds (Camel, Cavalier), made essentially the same case, adding: "One of the best ways of getting publicity is for a doctor to make some startling claim relative to people's health regardless of whether such statements are based on fact or theory." Another executive told *Business Week* (anonymously), "If we are guilty and they find out what causes cancer, we'll remove it from cigarettes." But no one asked him what would be done if cigarettes were established as guilty without it being clear which chemicals might be removed to make them safe.

Reassurance through advertising

What would master-advertiser George Washington Hill have done if he were around? The grim humor circulating at the time was "Hill would have known what to do about this health business. He would have made cancer fashionable."

During the '50s, the cigarette industry used a wide variety of messages intended to minimize the "health scare." Philip Morris advertised, "STOP WORRYING about cigarette irritation" and offered "THE CIGARETTE THAT TAKES THE *FEAR* OUT OF SMOKING!" Pall Mall promised to "guard against throat scratch." Camels assured that "More Doctors Smoke Camels." Camel tried the "healthy American" approach by showing an obviously un-cancerous group of athletes. Chesterfield ads promising "all the benefits of 30 years of scientific tobacco research," showed a spotless, gleaming North Carolina laboratory with brilliant-appearing young scientists extracting impurities from mysterious test tubes. Some ads soft-pedaled the health issue and emphasized the "pure pleasure" of smoking. Pall Mall, for example, suggested, "Reward Yourself."

By 1955, a research organization noted that belief in cigarette advertising ran from 4 to 8 percent, compared to 25 to 30 percent for refrigerator commercials. However, Americans still bought cigarettes.

In 1958, executives from the coffin-nail industry gathered for a press conference at New York's Plaza Hotel to launch the new "Hi Fi" filter Parliament cigarette. The *Nation* reported:

> In the foyers, test tubes bubbled and glassed-in machines smoked cigarettes by means of tubes. Men and women in long white laboratory coats bustled about and stood ready to answer any questions. Inside, a Philip Morris executive told the audience of reporters that the new Hi Fi filter was an event of "irrevocable significance." The new filter was described as "hospital white."

Two companies attempted to use the "health scare" to advantage. Using a "let's face it" technique, Rothman Ltd. of Canada in 1957 took a full page ad in a Canadian newspaper "accepting the *statistical* evidence linking lung cancer to heavy smoking . . . as a precautionary measure in the interest of the smokers," but suggesting that smokers turn to their "safer" cigarette, filter-tip Pall Mall. Later, a new American cigarette, king-sized Diplomat, was advertised as "safer" because a new method of curing supposedly reduced the toxic effects of tars and nicotine and lowered the combustion rate and temperature of the smoke.

Philip Morris became the first cigarette company to concentrate on the college market. The company distributed to hundreds of college newspapers a snappy and entertaining column, "On Campus with Max Shulman," which soon became steady reading for over a million students, professors and others. Campus sales were organized by full-time regional college supervisors who hired undergraduates as student representatives. The promotional techniques included competition between fraternities to guess football scores—with submissions written on the back of a Philip Morris wrapper. Prizes ranged from photographs and ping-pong games to trips to Europe. By 1960, Philip Morris had 165 campus representatives, some of whom graduated to key posts with the company.

And then there was the old standby, flag-wrapping. In an effort to keep America loyal to cigarettes, the Tobacco Institute sought to (and still does) glorify tobacco's role in American history. The between-the-lines message is that if you are anti-tobacco, you are anti-American. We were treated to complimentary copies of a beautifully illustrated historical treatise like TOBACCO AND AMERICANS. At one point, the Tobacco Institute promoted a gala celebration of tobacco's 350th year complete with a festival at Jamestown commemorating the shipment of the first tobacco crop from the Colony to England.

The most effective novelty of cigarette advertising in the '50s was the use of television. Beautiful girls, young virile men, athletes, and pilots smoked and told us of the supposedly good life with cigarettes.

In the late '50s, a survey of 1,100 top ad men indicated that more than half of them thought there was a definite or possible link between the smoking habit and cancer. About 30 percent of them had either cut down, stopped smoking completely or switched to cigars. But apparently 100 percent of those with tobacco accounts still kept pushing cigarettes.

Trouble with the FTC

The Federal Trade Commission monitored tobacco ads closely. In 1950 it strongly objected to a variety of claims involving health. Are cigarettes an aid to digestion? The FTC decided:

> The only physiological effect cigarette smoking can have upon digestion, if it has any at all, is harmful . . . such harmful effects may be interference with the normal gastric and intestinal motility, an increase in the acidity of the digestive fluids of the stomach, a lessening of the hunger sensation, or an aggravation of existing incipient gastrointestinal disorders.

Are cigarettes a pickup, a reliever of pain? The FTC ruled:

> The question whether or not smoking . . . accelerates the temporary release of existing bodily energy depends in large measure on the effect of such smoking upon the blood sugar level of the smoker. . . . In other words, no case can be made that cigarettes have any effect on fatigue.

A soother of nerves? An antidote for hair mussing and key jangling? No way:

> The smoking of cigarettes will not under any condition be physiologically beneficial to any of the bodily systems. . . . In the case of addicted persons, cigarettes can often afford the smoker some temporary relaxation. . . . [But] in the case of persons not accustomed to smoking, however, the effect of smoking even one cigarette will be just the opposite. Such a person will not only fail to have his nerves soothed or steadied, but he will probably become positively ill and quite upset as a result of his experience.

And to the question of whether cigarettes and athletes go together, the FTC responded:

> One cannot smoke as many cigarettes as he likes and keep in athletic condition because of the apparent adverse action upon the endurance and energy.

Toward the end of the decade, the FTC moved against Trim Cigarettes for suggesting that doctors approved of them as a weight loss aid.

More defensive tactics

History might designate Monday, January 4, 1954, as the day that the Tobacco Industry officially became dishonest. On that day, newspapers around the United States carried a full-page advertisement that began "A FRANK STATEMENT TO CIGARETTE SMOKERS." The ad claimed that:

1) Medical research of recent years indicates many causes of lung cancer.

2) There is no agreement among the authorities regarding the cause.

3) There is no proof that cigarette smoking is one of the causes.

4) Statistics purporting to link cigarette smoking with the disease could apply with equal force to any one of the many other aspects of modern life. Indeed, the validity of the statistics themselves is questioned by numerous scientists.

The companies who paid for the ad—all of the major firms except Liggett and Myers who thought it was a bad idea to get into debates with physicians and research scientists—announced formation of the Tobacco Industry Research Committee (TIRC), funded but supposedly not controlled by the tobacco industry. A few days later, it was announced that Dr. Clarence Cook Little, a geneticist and cancer specialist, had accepted the scientific directorship. Although he had been managing director of the American Society for the Control of Cancer (now the American Cancer Society), Dr. Little served as a vigorous spokesman for the tobacco interests. One of his most frequent lines was that the tobacco-disease link was "premature and oversimplified." He passed off the 18 major epidemiological studies demonstrating the link as "the opinion of a few statisticians." And he suggested that the problem lay not in tobacco, but in the type of individual who smokes. Perhaps something in the smoker's physical or emotional make-up causes a bad cancer risk with or without the cigarette habit.

The TIRC announcement and advertising campaign were a way of saying "We care about your health, American smoker," a message designed by Hill and Knowlton, the public relations firm that still works with the tobacco companies today. The companies did distribute unrestricted money for research—a fraction of their advertising budget—but they ignored the resultant negative medical findings!

The tobacco folks had a real scare in 1957, when Pope Pius XII suggested that the Jesuit order give up smoking. There were only 33,000 Jesuits in the world at that point, so the industry was not worried about losing this handful of smokers. They feared that the Pope or other church leaders might ask, as a magazine headline once put it "When are Cigs a Sin?" and worse yet, might answer "Always."

But the Industry has a way of turning things around. In *United States Tobacco Journal,* editor William G. Reddan noted that there has always been opposition to tobacco on moral grounds—from fanatics, but thank goodness the Pope is not a fanatic. The Jesuits have a way of life that "is traditionally stricter than other segments of the clergy or the laity in general," the editorial suggested. What the Pope was *really* saying was that cigarette smoking is fun, good, pleasurable and everything else wonderful; and the only reason the Jesuits should not partake is that they are supposed to flee from human gratification. Thus the Pope was actually endorsing tobacco for non-Jesuits!

The Tobacco Institute reacted more militantly to the so-called "Vanguard Issue." Vanguard was a tobaccoless smoke introduced in the Fall of 1959. The product's creator, Bantop Products Corporation of Bay Shore, Long Island, immediately ran into problems advertising it. In the New York metropolitan area, for example, only one newspaper would accept the ads, the others claiming that the "Now Smoke Without Fear" claim in the headline was unsubstantiated and disparaged the competition. Vanguard's makers claimed that the cigarette industry was conspiring to keep their tobaccoless product off the market. The tobacco industry admitted as much when the trade publication *Tobacco Leaf* editorialized:

> Under the circumstances the most effective weapon against invaders is economic pressure and we believe that it should be used in whatever legal manner the industry deems necessary for its own preservation.

In 1959, when the *Reader's Digest* carried a piece entitled "The Growing Horror of Lung Cancer," the tobacco industry was able to keep ads for the *Digest* article out of the New York transit system. F. Lee Moyne Page, president of Transportation Displays, Inc. which handled all rail commuter advertising around New York, noted that the manufacturers of Tareyton and Lucky Strike were among those who insisted that *Digest* advertising be removed. They wanted to replace mention of the lung cancer story with car cards for other *Digest* articles that month like "How to Act on Your Honeymoon" and "Being a Real Person." Mr. Page, quoted in *Advertising Age,* said, "This was done because of serious complaints from cigarette advertisers. Several tobacco companies directed our attention to the *Digest* copy as being hurtful to their interests, and we concurred."

The filter derby

For the first time in almost a quarter century the sales curve began leveling off. Total consumption for 1953 and 1954 declined over 6 percent and per capita consumption went down almost 9 percent. Cigarette consumption declined in 1953 to 423 billion from the 1952 record of 435 billion. The companies felt they had to do something. Their answer was to revive the old fashioned "mouthpiece" concept in the form of filtered cigarettes. Thus, while assuring smokers that there was nothing harmful about cigarettes, tobacco companies proceeded to develop supposedly safer ones. Filters had additional advantages—they eliminated loose tobacco ends and made cigarettes cheaper to manufacture because the inexpensive filter was a substantial portion of the cigarette.

Viceroy (Brown and Williamson) had had filter cigarettes on the market since the mid-1930s, but only after the "scare" of the '50s did they become popular. Soon there were many competitors. Reynolds had Winston, American offered Tareyton filters, Lorillard brought out Kent with the "micronite" filter so unique it was "developed by researchers in atomic energy plants." Actually, it turned out that the "micronite" filter worked too well. Smokers complained that all the "kick" from Kent was gone, and by 1954, Lorillard sneaked back to a looser filter that drew more easily and permitted additional tar and nicotine to enter the lungs. Not surprisingly, this "new feature" was not advertised.

In 1950 about 30 percent of cigarettes were filtered. By 1957 the figure was 50 percent. Filters presented a mixed bag for medical scientists. The filters would theoretically remove some of the nicotine and the products of combustion—tars—which might be harmful. But since no one really knew which ingredients in cigarettes were harmful, it was difficult to know if the filters were taking out the right substances. Beyond that was the concern that smokers would be misled into thinking that filtered cigarettes were safe.

Filters got a big boost from an unlikely source: *The Reader's Digest*. Looking at the trend toward filtered cigarettes, one *Digest* article put Kent on the map by discussing its "micronite filter." Soon thereafter began the "tar derby"—cigarette ads tripping over each other to proclaim which had the lowest tar and nicotine contents. The FTC had already banned use of words like "milder" and "smoother" in cigarette ad copy. In 1955 the agency prohibited either stated or implied medical approval of smoking. Unsubstantiated claims about nicotine, tars or other components were also forbidden.

The tar wars of the late '50s must have made many smokers helplessly confused. The term "high filtration" was introduced to indicate reduction of smoke solids. Philip Morris' new Parliament filter had "30,000 filaments." L&M miracle tips had United States patent number 2,805,671. Hit Parade's had "40,000 filter traps." Pall Mall declared that "Fine tobacco is its own best filter," and that its cigarettes would "travel the smoke further, to make it cooler and sweeter for you." Viceroy's "plus king size length" meant the "smoke is purified even more by the extra tobacco." Its new "Health Guard" filter, made of "estron," was said to be a "100 percent filter . . . Snow White, with 20,000 tiny filter traps." Tareyton touted a "genuine cork tip to protect the lips" and filter the smoke "naturally." Raleigh had no filters, just its coupons.

In 1959 the FTC informed manufacturers that "all representations of low or reduced tar or nicotine, whether by filtration or otherwise, will be construed as health claims. . . . Our purpose is to eliminate from cigarette advertising representations which in any way imply health benefit." This actually pleased the tobacco companies! No longer did they have to work on developing filters that were more effective than those of their competition.

Congressional action

What did the U.S. Congress do about cigarettes during the first decade following the flood of medical evidence against cigarettes? Not much. Except for the FTC's sometimes successful attempts to regulate advertising, not much *could* be done—for a number of reasons, including the fact that there was no official government stance on cigarettes and health.

In 1957, Representative John A. Blatnik (D-MN), then Chairman of the Legal and Monetary Subcommittee of the Government Operations Committee, conducted hearings to define more exactly the role and responsibility of the FTC regarding advertising claims for filter cigarettes. (Blatnik himself was a smoker until he underwent surgery a number of years later, and gave it up.)

The tobacco industry paraded its representatives (all smoking furiously) through the hearings to talk about the "controversy," and the "need for more data." Surgeon General Leroy Burney countered, "It is clear there is an increasing and consistent body of evidence that excessive cigarette smoking is one of the causative factors in lung cancer." Dr. John R. Heller, Director of the National Cancer Institute, who had been cautious in the early '50s, testified that the "overwhelming majority" of scientists and physicians in the Public Health Service now supported this position.

In 1958, the Blatnik committee report concluded, "The cigarette manufacturers have deceived the American public through their advertising of cigarettes . . . the FTC has failed in its statutory duty to prevent deceptive (practices) in filter cigarette advertising." Not long afterward, Blatnik's subcommittee was reorganized out of existence. (It was later revived, but due to the efforts of the tobacco lobby, Blatnik was not among its members.)

Another of the few government officials who took a leadership role on this issue during the '50s was Senator Richard L. Neuberger (D-Oregon). His favored target was cigarette price supports. In 1957 he introduced a bill to end price support marketing quotas, acreage allotments and acreage reserves for tobacco. He believed that taxpayers should not subsidize a cancer-causing substance. But we still do.

As the FTC scolded, Congressmen scratched their heads and medical researchers continued to accumulate evidence—Americans entered the '60s smoking more than ever. Sales had dropped after the "health scare" of 1954, but by 1959, it looked as if the tobacco industry had weathered the statistical storm. In 1960—nearly 70 million Americans still smoked—about the same number who voted in the 1960 Presidential election.

During the '50s, hundreds of thousands of Americans died of lung cancer, a disease almost unheard of 50 years previously. One of the victims was a physician involved in the kickoff of health data in 1950: Dr. Evarts Graham. He gave up cigarettes in 1953, but that was too late. Dr. Alton Ochsner described a letter from Dr. Graham as the "saddest letter I have ever gotten from anyone." Written two weeks before Graham's death in 1957, it stated:

> Because of your long friendship, you will be interested in knowing that they found that I have cancer in both my lungs. As you know I stopped smoking several years ago but after having smoked much as I did for years, too much damage had been done.

10
The Saga of the '60s

How the cigarette industry discovered its best filter yet: Congress

The overall reaction of Americans in the '50s to the growing conclusion that cigarettes were this country's leading cause of preventable death might well be characterized by the dictionary definition of *shock*: "a violent blow, shake, or jar; any sudden disturbance or agitation of the mind or emotions as through great loss or surprise."

The word to characterize the '60s might be *turmoil*: "a disturbance, tumult; confusion; uproar; commotion." This period saw the accumulation of even more unsettling medical data. It was a decade of official government resolution of the cigarette and health "controversy" with release of the 1964 Surgeon General's report, accompanied, of course, by the usual denials from the cigarette pushers.

Bureaucracy's many faces

During the '60s, government attitudes toward cigarettes ranged from support to apathy to opposition. Tobacco is one of the basic commodities covered by the Agricultural Adjustment Act of 1933. This meant that the tobacco industry, at government expense, was offered acreage restrictions and loans against surplus production. In 1939, Congress passed a law setting up an export corporation with government funds to buy surplus leaf on the domestic market to sell abroad. Thus, during the '60s, the government itself was in the tobacco business—at the same time that some government health officials were calling tobacco a hazardous substance!

In the early 1960s, Secretary of Agriculture Orville Freeman reasoned that price supports were beneficial. If the program were discontinued, he claimed, tobacco prices would fall, cigarettes would cost less and people would smoke more. In other words, tobacco subsidies were good for public health! Eager to increase sales of tobacco abroad, the Agriculture Department spent hundreds of thousands of dollars to advertise cigarettes to the people of Japan, Thailand and Austria. Uncle Sam also put up over $100,000 to produce a 23-minute promotional color movie called "The World of Pleasure," designed for free distribution to England, France, Belgium, Germany, Austria, the Netherlands, Denmark and the United Arab Republic—all, of course, translated into the appropriate language.

HEW Secretary Anthony Celebrezze thought that government should play no role whatsoever in advising the public of the supposed hazards of smoking. During mid-1964, Celebrezze informed the House Interstate and Foreign Commerce Committee that his department opposed all pending cigarette control bills.

The Public Health Service was interested in cigarettes, but in a rather limited way. In 1957, it had published a report concluding that, "The weight of the evidence at present implicates smoking as the principal etiological factor in the increased incidence of lung cancer." Senator Maurine Neuberger was disturbed by the weak tone of this statement and the fact that it came several years after other distinguished groups had ruled firmly against cigarettes. She was also concerned that the statement was limited to lung cancer even though it was known that smoking increased the risk of many other diseases. But the Public Health Service was under Mr. Celebrezze's Department of Health, Education and Welfare. So perhaps the agency had gone about as far as it could go politically.

What about the Food and Drug Administration, an agency one might think would want to regulate cigarettes? Perhaps taking its cue from the boss (Mr. Celebrezze), the FDA showed no interest in getting involved. Does it have jurisdiction over cigarettes? Senator Neuberger pointed out that tobacco was listed in the 1890 edition of the *U.S. Pharmacopoeia*, the government's official compendium of drugs. It was removed in 1905, she said, as the price paid to congressmen from tobacco states for supporting passage of the Food and Drug Act of 1906, which created the FDA. Eliminating the word "tobacco" prevented the substance from being regulated as a drug.

But if the FDA had wanted to get involved during the '60s, it probably could have done so. The Hazardous Substance Labeling Act, passed in 1960, included FDA jurisdiction over the sale of substances which can produce illness in man through inhalation. But as Senator Neuberger pointed out:

> The action—or inaction—of the Food and Drug Administration provides a fair sample of the overriding timidity and inertia that have plagued nearly every governmental response to the smoking problem. The Public Health Service, the Federal Trade Commission, the Department of Agriculture, Congress, and for the most part, the individual states and local governments have had a shared opportunity and obligation to aid in a constructive solution of the smoking problem. And each, to a greater or lesser degree, has failed rather dismally.

Perhaps she was a bit harsh on the FTC. In the early '60s it was itching to get into tobacco regulation, given its long and fairly successful history of monitoring cigarette advertising in previous decades. But this agency felt it could not move ahead with authority until there was some type of official government stance—or report—on the subject of smoking and health. That came in January, 1964.

Pressure mounts

In spring of 1961, the American Cancer Society, the American Public Health Association, the American Heart Association and the National Tuberculosis Association suggested by letter that a Presidential commission be appointed to study "the widespread implication of the tobacco problem." But President Kennedy ignored the request. Not long afterward, Senator Neuberger introduced Senate Joint Resolution 174 which called for the establishment of a Presidential Commission on Tobacco and Health. As she expected, the Senate ignored it.

Then an event occurred which seemed insignificant at the time but actually became a turning point. During the Presidential press conference of May 23, 1961, a reporter who had obviously been doing his homework asked President Kennedy what he thought the federal government should do about the growing concern about cigarettes. Mr. Kennedy replied, "That matter is sensitive enough and the stock market is in sufficient difficulty without my giving you an answer which is not based on complete information.... I would be glad to respond to that question in more detail next week."

Now on the spot, Kennedy referred the matter to Surgeon General Luther Terry for a response. Two weeks later, Dr. Terry (a cigarette smoker) announced, "It is timely to undertake a comprehensive review of all available data. I have therefore decided to appoint an expert advisory committee to study the evidence, evaluate it, and make ... recommendations." Actually, the available evidence against cigarette smoking was so overwhelming that the committee's verdict was predictable. The Surgeon General's committee simply represented a forum for the U.S. government to make an official statement on smoking and health.

Selected for the committee were Stanhope Bayne-Jones, M.D., LL.D., former Dean of Yale School of Medicine; Walter J. Burdette, M.D., Ph.D., Head of the Department of Surgery, University of Utah School of Medicine; William G. Cochran, M.A., Professor of Statistics, Harvard University; Emmanuel Farber, M.D., Ph.D., Chairman, Department of Pathology, University of Pittsburgh; Louis F. Feiser, Ph.D., Professor of Organic Chemistry, Harvard University; Jacob Furth, M.D., Professor of Pathology, Columbia University; John B. Hickam, M.D., Chairman, Department of Internal Medicine, University of Indiana; Charles LeMaistre,

M.D., Professor of Internal Medicine, The University of Texas Southwestern Medical School; Leonard M. Schuman, M.D., Professor of Epidemiology, University of Minnesota School of Public Health; and Maurice H. Seevers, M.D., Ph.D., Chairman, Department of Pharmacology, University of Michigan.

These individuals were among the most distinguished members of their professions. There were eight medical doctors, one chemist and one statistician. Three were cigarette smokers, and two smoked pipes and cigars on occasion. To guard against future accusations of bias, the Surgeon General included no one who had previously taken a stand on the tobacco-health question, and gave the tobacco industry, health groups, federal agencies and professional associations the opportunity to submit nominations—and to veto anyone they thought unacceptable.

Dr. Herman Kraybill, a distinguished scientist employed by the National Cancer Institute, had been named Executive Director of the Advisory Committee prior to the selection of the committee members themselves. Shortly after his appointment, when asked by a local reporter how he thought the verdict would go, Dr. Kraybill replied that the evidence "definitely suggests that tobacco is a health hazard." After this comment was published, he was removed from the committee.

The committee met on November 9, 1962 and held nine meetings of two to four days' duration during 1963. They held the meetings in absolute secrecy, shunned interviews, and generally conducted business as if they were working on an atomic bomb (which, in a manner of speaking, they were). *Newsweek* described the scene thus: "Outside the autumn sun played on the glass roof of the new National Library of Medicine in suburban Washington. In the windowless office deep in the basement, ten men quietly struggled through a white mountain of paper."

Astute reporters got a clue to the direction of the report when they saw Dr. Terry around Washington soon after the Committee began its work. He had a pipe, not a cigarette, in his mouth. About all the tobacco industry could hope for was a dissenting opinion or two.

A "preview" of sorts had taken place in England two years earlier. In 1962, the Royal College of Physicians in London had released a report titled "Smoking and Health," warning that, "Cigarette smoking is the most likely cause of the recent world-wide increase in deaths from lung cancer." The report suggested: 1) substituting pipe and cigar smoking for cigarettes; 2) discouraging smoking by adolescents; 3) restricting the advertising of cigarettes; 4) restricting smoking in public places; and 5) increasing taxes on tobacco. The British government endorsed the report, adopted educational measures designed toward discouraging smoking, and eventually did restrict advertising.

The Surgeon General's report

The official U.S. judgment on the cigarette was released early on Saturday, January 11, 1964—a day on which the stock market was closed. Behind locked doors, and surrounded by "No Smoking" signs, 125 reporters heard the grim warnings of the country's chief physician and his 10-member panel. Reporters were given 90 minutes to ask questions and scan the 387-page report. Apparently, they were held captive to minimize the possibility of leaving with half-baked versions of what was being said.

The Surgeon General's experts had concluded that smoking is causally related to lung cancer in men, outweighing all other factors, including air pollution. Information on smoking and lung cancer in women was not fully available because women had begun smoking in substantial numbers only 20 years before. But it seemed to the panel that the evidence pointed in the same general direction for women. The report also indicted cigarette smoking as a major cause of heart disease, bronchitis, emphysema, and cancer of the larynx. As for filter-tipped cigarettes, which accounted for 55 percent of the market, the committee found insufficient evidence that they did any good.

The bottom line was indeed bad news: *cigarette smoking was a health hazard of sufficient importance to warrant appropriate remedial action.* The only good news was that smokers could reduce their risks by giving up the habit.

What should the government do? Surgeon General Terry called for an "era of action" to discourage smoking. He suggested four possible government remedies: an educational campaign; labels on cigarette packs stating ingredients; stamped warnings about health hazards on packs and restrictions on advertising copy. Dr. Terry halted free distribution of cigarettes to the 16 public hospitals and 50 Indian hospitals under the direction of the Public Health Service; and he ordered the staff members of those institutions to begin educational programs to encourage people to quit smoking.

All smoking members of the Surgeon General's committee except 66-year-old Dr. Feiser gave up cigarettes. Two years after signing the Report—agreeing with every bit of it, but unable to give up his 45-year habit—Dr. Feiser underwent an operation for lung cancer which he described as having been "brought on by heavy smoking." After the operation, he told *Newsweek*:

> When we were working on the report, I was convinced about the findings, but I thought I was healthier than other people involved in the report and I also thought I was old enough so that if I was going to get it, I would have already had it. I was sure that this couldn't happen to me. After all, statistics are cold things. It's quite a different thing when it becomes a personal matter.

Tobacco industry reaction

Since making their first official public statement in 1954, the tobacco interests, now united through the Tobacco Institute, were consistent in their choruses of "Not proven." To suggest that the scientific community was still split on the issue of smoking and health, industry documents displayed a logo of scales in equal balance. (A more accurate representation would have been a 100-pound cement block on one scale and a grain of sand in the other.)

The Tobacco Institute was ready for the Surgeon General's report. That same afternoon they had press releases ready. "This report is not the final chapter," said George V. Allen, Institute spokesman, "I endorse [the call] for more research." Howard Cullman, President of the Tobacco Merchants Association and a Director of the Philip Morris Company, said, "We don't accept the idea that there are any harmful agents in tobacco." (The tobacco industry still mouths these opinions, even calling the 1982 Surgeon General's report "inconclusive.")

President Johnson had a number of opportunities to lend the prestige of the White House to the Report. But he didn't. In his Health Message of February 10, 1964—when the words of the Surgeon General were still fresh in the minds of Americans—he ranged over a wide variety of health problems from medical care for the aged to narcotics and drug abuse—but said not one word about tobacco.

On March 8, 1964, in reply to a question on this subject at a press conference the President noted: "I don't think that the Report has been made a government report as yet." Later that month he discussed his personal experience with cigarettes on a television interview. He had given them up on his doctor's advice after suffering a heart attack in 1955 and he told viewers: "I've missed it every day." Eventually he went back to cigarette smoking and died of a massive coronary attack in 1972.

Reaction from smokers

The 1964 Surgeon General's Report had an immediate negative impact on cigarette sales. Several months after the report was issued, the industry conceded that cigarette sales declined almost 20 percent for the first two months, but then returned gradually to their previous levels. *Science* magazine commented:

> The rise in consumption can in very large part be attributed to nothing more than the fact that some 70 million Americans find tobacco delicious to use and painful to discard; but a fair amount of credit for the restoration of sales must necessarily go to the tobacco industry, which had handled its peculiar problem with extreme shrewdness.

"Shrewdness" here is defined as heavy advertising to get people to forget quickly what the Surgeon General said. Many smokers reacted like the man in the story who had just seen a film on the removal of a cancerous lung: "After I saw that, I decided to give up going to the movies."

CIG vs. FTC

The Surgeon General's report had given the Federal Trade Commission what it needed to move ahead. One week after the Surgeon General's press conference, the FTC proposed new rules to govern the advertising and labeling of cigarettes. As of January 1, 1965, all cigarette packages would have to carry a warning that smoking may cause death from cancer and other disease. Public hearings were scheduled. With that, the ears of Congress and the tobacco industry perked up.

The Commission said it was concerned with two ways that cigarette advertising might be unlawfully misrepresenting or concealing the health hazards of smoking. First, the ads gave the false impression that smoking promoted health—or at least presented no health hazard. Second, the ads created a psychological and social barrier to public understanding of the seriousness of the problem involved. The agency noted:

> Massive advertising, depicting and constantly reiterating the pleasures and desirability of cigarette smoking but failing to disclose the risks to health, appears to be a potent force in increasing the sale of cigarettes, despite increasing scientific and government recognition of the existence and seriousness of such perils.

The FTC proposed two remedies: a dramatic warning label on the package, and a similarly attention-grabbing label in all cigarette advertising. Surgeon General Terry supported the rulemaking procedure because he was "convinced that the American people have been deceived and misled by cigarette advertising—and their health has been harmed as a consequence." The tobacco industry countered by questioning the FTC's authority to make policy, and worked to insure that the final policy decision would be made by Congress, not the FTC.

Members of Congress from tobacco states were especially powerful at that time. In the Senate, nearly one-fourth of the committees were chaired by men from the six tobacco states. The tobacco industry is located almost entirely in the southern part of this country where one-party dominance provides high congressional seniority. Given the power distribution in Congress, the members from the tobacco states could exert a tremendous influence on matters that came before their committees. And they did. Soon after the FTC's announcement, Congress announced hearings of its own.

The FTC hearings began in March, 1964. The industry brought up every imaginable argument about why the FTC should not be involved with cigarettes. They said the agency was exceeding its statutory authority by attempting to do so, that the industry should be permitted to police itself, and that if any restrictions were to be put on them, they should come only from the U.S. Congress.

The tobacco industry got some curious support from the American Medical Associaton. Executive Vice-President F.J.L. Blasingame noted that the AMA had "historically endorsed and promoted federal and state legislation containing labeling requirements with respect to the sale of drugs, cosmetics and hazardous household products to the consumer." But he went on to state:

> With respect to cigarettes, cautionary labeling cannot be anticipated to serve the public interest with any particular degree of success. The health hazards of excessive smoking have been well publicized for more than 10 years and are common knowledge. Labeling will not alert even the younger cigarette smoker to any risks of which he may or may not be already aware.

Why did the American Medical Association join forces with the tobacco people on this issue? Even if warning labels were not proven effective, why would the AMA object to them as a first step in public education about cigarettes? Cynics suggested that AMA representatives were attempting to appease Southern congressmen to ensure their support against socialized medicine—a charge that the AMA termed "ridiculous."

Actually, the AMA's involvement—or lack of it—in the tobacco issue during the '60s is quite puzzling. In December 1963—just a few weeks before the Surgeon General's Report was released—the AMA's House of Delegates approved a broad program of research into questions relating to smoking and health. Of course, this pleased the tobacco folks. "We are gratified," said Dr. Clarence Cook Little, one of the tobacco spokesmen, "by the reports of the AMA's recognition of the need for additional research on smoking and health." The research program was supported by a $10 million grant from the six largest tobacco companies!

In 1964, the AMA did publish an educational leaflet announcing that smoking is a "threat to life." Why? Because "numerous deaths occur each year from burns and suffocation due to falling asleep while smoking." Under the heading, "Suspected Health Hazards," the AMA flyer concluded that according to *some* researchers, cigarette smoking "shortens life expectancy" and is "alleged to cause cancer of the lungs and bladder." But the leaflet went on to assure us, "Some equally competent physicians and research personnel are less sure of the effect of cigarette smoking on health . . . Smoke if you feel you should, but be moderate."

During the summer of 1964, about two months after the public hearing, the Federal Trade Commission ruled that henceforth it would be an "unfair or deceptive practice" for any manufacturer "to fail to disclose clearly and prominently in all advertising and on every pack, box, carton or other container in which cigarettes are sold . . . that cigarette smoking is dangerous to health and may cause death from cancer and other diseases." The ruling was supposed to take effect on January 1, 1965. But it didn't.

Ironically, the FTC's ruling made no one happy. Health advocates felt it was too weak, and cigarette companies were especially disturbed by the prospect of warning labels in advertising. By the time the ruling was announced, however, they were well equipped to help themselves out of this jam.

The tobacco industry reacts

Late in 1963, Philip Morris had engaged the Washington law firm of Arnold, Fortas and Porter. The firm's senior partner, Abe Fortas, a friend and confident of President Johnson, was chosen by the six major tobacco companies to form a committee of lawyers to solidify industry togetherness. The committee met almost daily, planning for every possible contingency, and carefully forming the industry argument for the FTC hearings. When the issue of labeling came before Congress, it was this group who wrote the testimony, conducted the search for friendly witnesses, and even supplied questions that its Congressional allies could ask opposing witnesses.

In another astute move, the six major companies (R.J. Reynolds, American Tobacco, Brown and Williamson, Liggett and Myers, P. Lorillard and Philip Morris) chose a lobbyist. Not any old lobbyist mind you, but Earl C. Clements who had represented Kentucky in both the House and the Senate (which gave him, as an alumnus, floor privileges), and who had very close personal and political ties with the President. Clements was the No. 2 man in the Democratic majority when Johnson was Senate majority leader and had served as Johnson's campaign coordinator for the 1960 election. His daughter, Bess Abell, was secretary to Mrs. Johnson, and his son-in-law, Tyler Abell, had recently been appointed by Johnson to serve as an assistant postmaster general. Clements not only had excellent political connections; he was well liked, gentlemanly, and a shrewd master of legislative infighting.

Clements quickly persuaded the House Interstate and Foreign Commerce Committee to ask the FTC to postpone implementation of their rule for six months. The FTC agreed. Hearings were then scheduled before the House committee and the Senate Commerce Committee where Clements had many good friends. One member of the House panel was Horace Kornegay of North Carolina who later left Congress to become chairman of the Tobacco Institute. Seven other members of the 33-member committee came from tobacco-growing regions.

When hearings began, Kornegay brought in a potted tobacco plant and declared, "This tobacco plant stands as the defendant in this trial." Then he emphasized the contribution of the tobacco industry to the economy of the nation and claimed that if legislation inhibiting the cigarette in any way were passed, thousands of jobs would be lost. (Congressmen are particularly sensitive to such economic shock waves which might directly affect their constituents and cause them to pressure their representative.) Clements had advised tobacco representatives to play up the economics and longstanding history of tobacco in America—and stay as far from the health issue as they could. In line with this, an industry press release stated, "Tobacco products pass across more sales counters more frequently than anything except money."

Tobacco's political coalition still includes growers, manufacturers, and distributors, and indirectly a clientele of millions who smoke billions of dollars of tobacco products annually. The recipients of those billions include manufacturers, advertising agencies, the mass media, farmers, shopkeepers and tax collectors. In *Smoking and Politics*, A. Lee Fritschler calls this "the tobacco subsystem." Obviously, any political challenge to cigarettes is fraught with difficulties.

Self-regulation?

Just before the hearings began, the cigarette industry declared that it would start regulating itself. This decision may have been prompted by Leroy Collins, President of the National Associaton of Broadcasters. Speaking at a 1963 conference of the NAB, Collins took the industry by surprise by accusing it of irresponsibility regarding advertising. He called on the broadcasting industry, which received the largest portion of the advertising pie, to take corrective action against certain kinds of television commercials directed at children. Collins, a reformed smoker, stirred up a storm when he added: "Tobacco advertising having a special appeal to minors, expressed or implied, should be avoided."

For a time it looked like Mr. Collins would be rewarded for his advice by being asked to join the world of the unemployed. But industry leaders were shrewd enough to realize that if they didn't at least pretend to discipline themselves, the government would. In 1964, the industry announced that it would embark on "self-regulation" in cigarette advertising, to cut the appeal of ads to children and to stop saying or implying that smoking is good for health. No longer would there be testimonials by noted athletes (although they planned to go on sponsoring sports events, as they still do today); the models that paraded before the camera with a cigarette would have to be at least 25 years old; and there would be no advertising on programs "primarily" aimed at children. This appeared to be a sort of confession of past sins and a promise to clean up the act. Regulations were drawn up by the Fortas legal group. Violation was to be punishable by a fine up to $100,000 at the discretion of the industry-appointed administrator, Robert B. Meyner, former governor of New Jersey.

The 1965 labeling decision

The hearings before the House and Senate committees during 1965 were similar. Both sides had expert witnesses. (Nearly two years after the hearings it was disclosed that a few of the witnesses for the tobacco interests had not properly identified themselves as paid industry consultants.) There were physicians who told of the health hazards of smoking—and industry employees who sang the glories of tobacco. Lobbyist David Cohen of the Americans for Democratic Action characterized the contest between the tobacco interests and the pro-health groups as similar to a match between the Green Bay Packers and a high school football team. It was also said that the only difference between the success in eradicating malaria and the failure to eradicate tobacco-related disease was the fact that the mosquitoes did not organize in their own defense. The tobacco bugs did so and quite successfully.

Senator Maurine Neuberger—long known as an advocate of tighter federal regulation of cigarette merchandising—had recently won a seat on the Senate Commerce Committee. To make the industry's task more difficult, she introduced legislation to give Congressional approval to the FTC's proposed rules. But she was ultimately alone on that issue.

Throughout the hearings, the tobacco spokespersons carefully claimed that medical opinion was still divided over the Surgeon General's report. They also suggested that what Mrs. Neuberger and the FTC wanted to do was "extremist" and could wreak havoc (here comes the American flag again) with a basic American industry "as old as Jamestown." They acknowledged, yes, that "heavy" or "excessive" smoking could be hazardous—but so were excesses of everything else. Would Congress want to put warnings of possible death on liquor, food and automobiles? Where would it all end? This confuse-them-by-arguing-to-the-point-of-absurdity routine overlooked the fact that while all of these other items could be used safely, *cigarettes could not*.

In the end, the House Committee recommended a measure to require health warnings on cigarette packs, but not in advertising. At the same time, it agreed to bar the FTC from acting on its own. Eventually the Cigarette Labeling and Advertising Act passed both houses of Congress and was signed into law in July 1965.

The law said that beginning on January 1, 1966, all cigarette packages must carry the warning, "Caution—Cigarette Smoking May Be Hazardous To Your Health." But, as part of the deal, the law forbade the FTC from even considering a warning in advertising until July 1, 1969.

To the naive, it may have appeared that the pro-health forces had won a Congressional victory. But as Elizabeth Brenner Drew put it in the September 1965 *Atlantic Monthly*, "The bill is not, as its sponsors suggested, an example of Congressional initiative to protect public health; it is an unabashed act to protect private industry from government regulation." Her article was titled, "The Quiet Victory of the Cigarette Lobby: How it Found the Best Filter Yet—Congress." Columnist Drew Pearson agreed: "The Congressional steamroller was never in higher gear than when the tobacco lobby rammed the so-called cigarette labeling bill through the house." He termed it "a shocking piece of special interest legislation ... a bill to protect the economic health of the tobacco industry by freeing it of proper regulation."

The tobacco interests were grinning from ear to ear. Why shouldn't they? Compare what almost happened to what actually did. Clements had told his clients that *something* was going to happen, either at the FTC or in Congress. In the light of the Surgeon General's report, the mood of the country would not let cigarettes completely off the regulatory hook. Pro-health forces wanted a strong label on cigarettes, noting specific links with diseases like cancer. Instead, they got a wishy-washy "May Be Hazardous." Pro-health forces wanted the health warning intimately tied to advertising to counterbalance the glamor, appeal and alleged "healthiness" of smoking. That was the *last* thing the tobacco companies wanted. Advertising had *made* the cigarette—and advertising mixed with scary talk about getting sick could *kill* the cigarette. The tobacco people felt so strongly about avoiding warnings on television ads that they eventually chose to remove their own ads completely rather than accept a warning.

During the early 1960s, state and local governments also began to consider cigarette regulation. For example, package warnings had been proposed for New York City by the health commissioner, and similar suggestions were being considered by state legislators in New York, Massachusetts and elsewhere. The governor of California had created a cigarette smoking advisory committee. If there was anything the cigarette companies wanted less than federal regulation, it was state or local regulation. The chaos which that could cause to marketing and sales was obvious. The Cigarette Labeling and Advertising Act neatly prevented this by making regulations uniform in all states. As summarized by Senator Frank E. Moss (D-Utah) in 1969: "In exchange for eleven words on the side of the cigarette package, Congress exempted the cigarette industry from the normal regulatory process of federal, state and local regulations." Other observers referred to the new law as "the rape of 1965."

In retrospect, it is clear that the industry came out way ahead. It not only kept government regulation to a minimum, but it also got an extra that would come in handy in court. With a label on the package, it would be difficult if not impossible for consumers to claim ignorance of the hazards of smoking. Thus, it could be argued that as of January 1966, smokers were using cigarettes at their own risk.

11

Showdown in Marlboro Country 1968-1970

Executive #1: Let's guarantee that every pack of Isano cigarettes is definitely less harmful than any other pack of cigarettes on the market.
Executive #2: How can this be substantiated?
Executive #1: Very simple; we put only 19 cigarettes in every pack. From a skit presented by NBC to its affiliates; as reported in *The New Republic*, May 9, 1964

During the 1960s, the tobacco companies formed one of the most powerful lobbying groups in American history. Their strategy was carefully planned, guided by the best expertise available. Their methods were simple: to deny any association between ill health and cigarette smoking; to distract American smokers from the enormous mound of data confirming cigarette hazards; and to seduce new and younger customers.

Propaganda tactics

During this 10-year period, tobacco companies played up the views of dissenting doctors. Although the medical verdict on cigarettes was clear, a few dissenting physicians provided the Tobacco Institute with marvelous copy for press releases.

One example was a 1963 article in *California Medicine* by Leroy Hyde, M.D., of the Pulmonary Disease Service at the Veterans Administration Hospital in Long Beach, California. After asking, "What do these statements have in common?" Dr. Hyde listed some old-time medical fallacies: Pellagra is an infection; blood passes from right to left side of the heart through invisible pores; and a good treatment for tuberculosis is horse riding. And then he tells us we can add to the list the statement that cigarette smoking causes cancer. These comparisons, of course, are quite unfair. The fallacies to which he referred were speculations that arose when medical science was in its infancy. By 1963 the case against cigarette smoking was *overwhelmingly* supported by scientific studies.

The Mayo Clinic's Joseph Berkson, M.D., regarded as the dean of American medical statisticians, regularly expressed his doubts that cigarettes caused lung cancer—and the Tobacco Institute saw that they were well publicized.

In 1967, Hawthorn Press published *It is Safe to Smoke*, by Lloyd Mallan. Noting that the book left him coughing after the third page, Morton Mintz summed up his views in a *New Republic* review titled, "It's Not Safe to Read." The book ran into trouble when some people complained that it read like an ad for Liggett and Myers' Lark cigarettes. After the FTC began to investigate the possibility that the industry had subsidized the book, Hawthorn took it off the market.

And as part of their PR campaign to confuse, the tobacco industry even sponsored their own studies. In a 1962 issue of *Industrial Medicine and Surgery*, Jacob Cohen, Ph.D., of New York University and Robert K. Heimann, Ph.D., from the American Tobacco Company, concluded that heavy smoking and low mortality went together in their "14½-year test of the Cigarette Hypothesis of Lung Cancer Causation," a poorly designed study which yielded results diametrically opposed to all other research on the effects of smoking on the incidence of lung cancer.

The case of the smoking dogs

Tobacco companies also claimed that the case against cigarettes is not closed because "animals don't get cancer." Researchers had long known that skin cancer could be produced by painting the skin of mice and other animals with tar condensed from cigarette smoke. But until the late 1960s there was no direct evidence that cigarette smoking could cause cancer in animals. The reason for this is that animals—perhaps because of some innate wisdom—do not voluntarily smoke.

In 1963, Dr. William G. Cahan of Memorial Sloan-Kettering Cancer Center devised a method of inserting a plastic tube through an opening in beagles' windpipes and pumping in smoke drawn from a cigarette. Various other devices were used to pump smoke into the dogs' lungs. Soon they seemed to like it. In 1966, Drs. Oscar Auerbach and E. Cuyler Hammond announced that they had found cell changes which resembled cancer in the bronchial linings of five beagles who smoked for 420 days. Emphysema and other lung diseases also developed. At a news conference in 1970, the researchers announced that their dogs who smoked for longer than 420 days developed lung cancer. It was also noted that filtering the smoke had reduced the incidence of cancer and that the degree of lung damage had progressed with the duration of smoking.

For many years, the tobacco industry maintained that the case against cigarettes could not be proven without showing that smoking could cause cancer in animals. But after the dog studies were published, the industry changed its mind, claiming it was impossible to "draw meaningful parallels between human smoking and dogs subjected to these most stressful conditions."

Actually, the parallel was right on target! In 1957 Dr. Auerbach and colleagues reported a well-designed study of autopsy examinations of 402 males to determine the incidence of cancerous and precancerous changes in the cells lining the respiratory tract. Precancerous changes were absent in nonsmokers and rare in light smokers, but were present in 4.3 percent of specimens from those who smoked 1-2 packs per day and in 11.4 percent of those who smoked 2 or more packs daily. Subsequent studies have confirmed this relationship for female smokers as well as males.

Irresponsible advertising

A 1964 survey by *Printer's Ink* found that ad men were unsure about the safety of smoking. Thirty-five percent of them admitted to stopping smoking, and 15 percent had cut down. However, not one of the 20 cigarette copywriters interviewed admitted to any pangs of conscience at creating copy for a product reported by the government to be harmful. One writer said, "Writing cigarette copy doesn't bother me one bit! Why should it? Should an automobile copywriter worry about writing copy for automobiles just because thousands of people die in car accidents every year?" Another writer said, "It's all a fad. A little while back it was the scare over cholesterol and heart disease." And a third said, "I write for the man who feeds me." With hundreds of millions of dollars being poured into cigarette advertising, the warning on the side of the packages was little more than a required bunch of letters. Thomas Whiteside's book on advertising, *Selling Death,* was aptly named.

The ethics of cigarette advertising of that period was perhaps illustrated best by the Kent ad which claimed that "No medical evidence or scientific endorsement has proved any other cigarette to be superior to Kent."

Another company offered a new twist on the old coupon promotion technique. If you bought enough cigarettes you could earn a new Autumn Haze Mink Stole. The company was kind enough to provide—upon request—a starter kit of 100 coupons. Then all you needed was to smoke two packs a day for 46 years to acquire the additional 47,185 coupons needed. (Or you and your spouse could join forces for 23 years.)

More government action

The Public Health Service continued to assign resources to the National Clearinghouse for Smoking and Health. The PHS also appointed two separate panels. The first, composed of 14 experts from various fields, reviewed the medical research and reported that the scientific evidence suggested that lowering the tar and nicotine content of cigarettes could render these effects less harmful. As a result, the FTC reversed its ban on the use of tar and nicotine figures in ads and, after setting up a standardized system for measuring these substances, encouraged manufacturers to use these data in their ads.

Cigarette companies have mixed feelings about using references to tar and nicotine in their ads. On one hand, it gives them a point of competition for whoever is lowest. On the other hand, such references call attention to the unpleasant topic of health hazards. Beyond that, the less tar and nicotine, the worse a cigarette "tastes"—which may cause sales to drop.

In 1967, the Surgeon General set up another panel—the Task Force on Smoking and Health—which included baseball star Jackie Robinson and former ad man Emerson Foote. The panel's conclusions were of the good news/bad news variety: cigarette sales appeared to be leveling off, but the ads were still misleading. The Task Force asked the government to take some action to balance the claims and appealing imagery in ads with the harsh realities about the health effects of smoking. The panel recommendations not only perked up the ears of the tobacco industry but made them angry. Like a wounded animal, the industry cried this report "is a shockingly intemperate defamation of an industry which had led the way in medical research to seek answers in the cigarette controversy."

HEW Secretary Wilbur J. Cohen, breaking the tradition of timidity set by a number of his predecessors, then complained to Congress that "The remedial action taken until now has not been adequate." His letter was accompanied by a lengthy bibliography and analysis of the hundreds of medical articles and research studies published since the 1964 Surgeon General's report.

The FTC, apparently recovered from its 1965 Congressional rebuke, submitted more recommendations. The first set, in 1967, suggested that a warning statement be required in all advertising and that the statement itself, then appearing only on packages, be made stronger: "WARNING: CIGARETTE SMOKING IS DANGEROUS TO HEALTH AND MAY CAUSE DEATH FROM CANCER AND OTHER DISEASES." A second FTC report was even bolder. It contained some analyses of advertising—and an example (the *True* magazine incident described below) of the tobacco interests' efforts to mislead smokers about the relationship of cigarettes and health.

Clearly, a consensus was emerging in both Congress and the regulatory agencies that cigarette advertising was getting out of hand. The tobacco companies insisted then, as they still do, that their ads were not meant to attract new smokers, but only to get current smokers to switch brands. They claimed they were *not* going after the youthful market. Some ad copy even described smoking as an "adult pleasure." But this argument is hollow. Ads portraying the cigarette as a sign of adulthood were obviously designed to attract young people to the habit. For example, American Tobacco began advertising in 1962 that, "Smoking Is a Pleasure Meant for Adults," but added, "Lucky Strikes Separates the Men from the Boys . . . but not from the Girls," which is just what the young boy trying to "prove" he was a man wanted to hear. The tobacco industry's contention that it wasn't trying to attract new smokers made no business sense at all. How could it prosper without them?

A planted story backfires

One particular move by the industry irked the FTC and just about everyone else who found out about it. The January 15, 1968 issue of *True* magazine contained an article by Stanley Frank called, "To Smoke or not to smoke—that is *still* the Question." The article dismissed the evidence against smoking as "inconclusive and inaccurate" and claimed that "Statistics alone link cigarettes with lung cancer . . . it is not accepted as scientific proof of the cause and effect." A few months later a similar but shorter article appeared in the *National Enquirer* entitled "Cigarette Cancer Link is Bunk" written by "Charles Golden" (a fictitious name commonly used by the Enquirer). The real author was Stanley Frank.

Two million reprints of the *True* article were distributed to physicians, scientists, journalists, government officials, and other opinion leaders with a little card that said, "As a leader in your profession and community, you will be interested in reading this story from the January issue of *True Magazine* about one of today's controversial issues." The tab for all of this was picked up by Brown and Williamson, Philip Morris and R.J. Reynolds. Later it was shown that author Frank had been paid $500 to write the article by Joseph Field, a public relations man working for Brown and Williamson, and then B&W reimbursed Field for this amount.

A great fuss was made about the *True* article. The FTC investigated the circumstances of its creation, the Surgeon General spoke out against its scientific errors, and *Consumer Reports* carried a full-length article about the incident in its June 1968 issue. Many industries plant stories which promote their cause. The most important part of the *True* story is not the fact that the tobacco industry did it—but that they were apparently under so much pressure that they had to use such a desperate measure. *Advertising Age* made this comment in an editorial:

> There is a vast difference between views that are clearly labeled, so that anyone reading them can ascertain for himself if the party or parties expressing them represent one faction or another, and views that are passed along as unbiased editorial comment "signed" by the editors of the magazine.

Enter the FCC

The growing concern was there, but as so often happens, it took the action of one individual to turn the tide. In this case, the individual was John F. Banzhaf, III, who, one day in 1968, saw one too many cigarette ads. Banzhaf, then working for a law firm which represented Philip Morris, wrote to WCBS-TV in New York City asking for free air time to respond to the cigarette commercials. When the station refused, Banzhaf wrote more letters to the station and to the Federal Communications Commission (FCC) as well. He struck out with CBS, but struck oil with the FCC.

In 1967, the Commission ruled that the "fairness doctrine is applicable to such advertisements." Banzhaf didn't get "equal time" because, as the FCC put it, "The practical result of any roughly one-to-one correlation would probably be either the elimination or substantial curtailment of broadcast cigarette advertising." But the FCC did support the request for "a significant amount of time for the other viewpoint," thus assuring a "grant" to pro-health activists of tens of millions of dollars of free TV time.

The Commission's decision sent shock waves through not only the tobacco industry, but also the hierarchy of the National Association of Broadcasters. The NAB called it an "unwarranted and dangerous intrusion into American business."

"Do we want the FCC to be able to prohibit the advertising of milk, eggs, butter and ice cream on TV?" cried North Carolina Democrat L. H. Fountain. "What if a group claims that automobiles are unsafe and candy and soda rot your teeth?" asked one network official. "Where does it stop?" "What if a woman objects to our girdle ads," despaired a tobacco industry attorney, "Does she get a chance to respond too?" The FCC made it very clear in their letter that they were focusing on a unique situation, cigarettes, and did not intend to take on any other products. Congressman Walter Jones (D-NC) also tried to rally support against the FCC from other industries whose products he claimed might face a similar dilemma in the future, but he was unsuccessful.

After the FCC denied the petitions for reconsideration, the National Association of Broadcasters filed an appeal in—cigarette country, of course—Richmond, Virginia. But Banzhaf had beaten them to the punch by filing his own appeal in the District of Columbia on the basis that his request for exactly equal time had been turned down by the FCC. As he expected, Banzhaf "lost" *that* appeal, but the court's decision on November 21, 1968 upheld the FCC's ruling that "significant amounts" of free time must be made available for anti-smoking commercials. "The danger cigarettes may pose to health is, among other things, a danger to life itself," the court wrote. The cigarette companies, demonstrating in this case that they would rather "fight than switch," took the matter to the

Supreme Court which eventually refused to review the case, leaving the appeals court decision standing.

While the legal questions were pending, the United States Public Health Service, the American Cancer Society and numerous other health organizations had prepared TV and radio spots to scare the stuffing out of smokers. When the question "Do you mind if I smoke?" was asked, the answer was always "yes." And a favorite and effective motto was "Cigarettes—they're killers."

But by far the most penetrating and effective anti-smoking ad was the one-minute spot featuring William Talman, a 3-pack-a-day man who had played Hamilton Berger, Perry Mason's legal opponent on television. Looking very ill—and under heavy sedation . . . he said (with reference to the fact that he had lost all 251 of his TV court cases to Mason),

> I didn't really mind losing those courtroom battles. But I'm in a battle right now I don't want to lose . . . I've got lung cancer. So take some advice about smoking and losing from someone who's been doing both for years. If you haven't smoked, don't start. If you do smoke, quit. Don't be a loser.

Talman's death at age 53 by the time the commercial was aired made the appeal even more dramatic, as did the background shots of the handsome young family he had left behind.

The American Cancer Society reported that in the 3½ years before the FCC applied its fairness doctrine to cigarette ads, it had distributed some 982 pre-recorded anti-smoking commercials to radio and television. During the eight months after the FCC's decision, the Society distributed 4,723 such commercials. These ads were effective and contributed to the reduction in cigarette consumption first noted in 1968. Government statistics indicated that as many as 10 million Americans quit smoking from 1967 to 1970. Inspired no doubt by his success, John Banzhaf founded Action for Smoking and Health, an anti-smoking educational and lobbying group based in Washington, D.C.

The FTC and Congress: Round #2

When the Cigarette Labeling and Advertising Act of 1965 was passed, no one really believed that the issue of cigarettes and health had been resolved. Pro-health forces were particularly frustrated by the fact that Congress had tied the hands of the FTC for four years. But as the deadline on FTC Congressionally-imposed inactivity neared, the FTC began flexing its muscles.

In February 1969, six of the seven Commissioners proposed a rule to prohibit cigarette advertising on radio and television. TV ads for cigarettes had been banned in England in 1965. Calling cigarette smoking "a most serious, unique danger to public health" the FTC concluded, "It would thus appear wholly at odds with the public interest for broadcasters to present advertising promoting the consumption of the product posing this unique danger—a danger measured in terms of an epidemic of deaths and disabilities." To head off any objection that they were setting a precedent, the Commissioners added that they were unaware of any other product commercials calling for such action and had no intention of proceeding against other product commercials.

The Tobacco Institute called the FCC decision "arbitrary in the extreme" and "an obvious threat to usurp the Congressional function." And Vincent T. Wasilewski, president of the National Association of Broadcasters protested, "Not only do we deplore the assumption of such power, but we deny that such power exists." Broadcasters, of course, were facing the possibility of losing nearly 10 percent of their total advertising revenues. In the past few years, cigarette companies had become their Number 1 client.

As was the case in 1965, the tobacco industry position was carefully and skillfully arranged, but by now, opposition to cigarettes had grown appreciably on Capitol Hill. The only staunch supporters of the industry remaining were those from the big tobacco states: the Carolinas, Kentucky, and Virginia.

The House of Representatives remained allied to tobacco interests, reporting a bill on June 18, 1969, to prohibit the states permanently and the federal agencies for six years from acting on cigarette advertising, in exchange for a slightly strengthened warning on the package. But before they did that, they had hearings which Chairman Harley O. Staggers described as the longest in his memory. Four fat volumes totaling 1686 pages appeared afterwards. Many of the witnesses—and much of the testimony—sounded the same as in 1965.

Again, the American Medical Association was conspicuously absent from the list of supporters. Since the AMA had contributed more than $600,000 to election campaigns in 1968, its view carried a political wallop. But it remained mute.

The *New England Journal of Medicine* said this about the House hearings:

> The transcript . . . makes appalling reading. Anti-smoking witnesses received perfunctory introductions and dismissals, were asked irrelevant or harrassing questions (. . . of Mr. John Banzhaf, "Are you primarily interested in the health of the people or are you primarily using this committee as a means of soliciting business?") and were exposed to the type of cross-examination that is used by hostile lawyers; to Surgeon General William Stewart, "I would like you to show me one instance of a laboratory report of definite causation of lung cancer by inhalation of cigarette smoke. Can you do that? Yes or No?"

By contrast, reception of the parade of pro-tobacco witnesses was friendly, if not ecstatic. "It is my privilege to welcome before the committee Joseph F. Cullman, III [Chairman of the Board and Chief Executive Officer at Philip Morris, Inc., as well as chairman of the executive committee of the Tobacco Institute]. He is head of one of the major industries in this country. He is an important business fixture in our area as well as an important civic leader. You will find him, I am sure, both knowledgeable and fair in this matter that is of great concern to us."

Industry legal counsel Thomas Austern had sent shock waves throughout the trade a few months before when he argued before the Federal Trade Commission that there was nothing deceptive about depicting smoking as a desirable habit without mentioning any possible hazard because "anybody who is not deaf and blind knows it is a hazard." Mr. Austern later explained his views did not necessarily reflect those of his clients on this issue.

Perhaps the tobacco companies thought they had overplayed their hand. Taking the Senate's pulse, they found that the time had come to pull back. It was clear that in the showdown in Marlboro Country, the Senate lawmen were going to outdraw them.

Michael Pertschuk, a young lawyer who was counsel to the Senate Commerce Committee, played an important role in getting the cigarette companies to surrender. Pertschuk coordinated the preparation of the Senate hearings on the House-passed bill. He showed the Senators 20 uninterrupted minutes of cigarette commercials stressing romance, pretty girls and athletes—all obviously pitched to young people. Another important figure in swaying the Senate was Senator Warren Magnuson, who by that time had expressed regret about the Congressional action in 1965. He showed little willingness to compromise this year. After Senator Vance Hartke of Indiana, who had been the chief spokesman for the cigarette lobby, defected, the tobacco industry was unable to find a single other Senator to represent their point of view on the Senate Commerce Committee.

Recognizing their inevitable defeat at the hands of the Senate, the tobacco interests folded. Joseph Cullman, III appeared before the subcommittee and announced that the cigarette manufacturers were willing to withdraw all radio and television advertising, beginning January 2, 1970, as long as Congress would extend to them antitrust immunity. The tobacco companies may well have been relieved that the end of *their* ads would sharply curtail anti-smoking ads which were terrifyingly effective. Congress agreed to the January 2nd cutoff and also strengthened the warning on the package: "Warning: The Surgeon General has determined that Cigarette Smoking is Dangerous to Your Health."

Why didn't the ads stop on the first of the year? Congress didn't want to deprive the tobacco men of their last licks in advertising during the New

Year's football games. And did they ever get those last licks! In a manner analogous to the final fireworks on the Fourth of July, the industry bombarded the airwaves on January 1st. Philip Morris spent $1.2 million for commercial time on all three networks between 11:30 P.M. and midnight alone—so that the cowboy from Marlboro could take his last ride on television.

Now for the postgame wrap-up. Who "won"? Newspapers and magazines heralded the news as a great step forward in public health. It was also a great step forward in improving their own economic health. Contrary to the promises of the cigarette industry during the hearings, the print media was about to be the proud recipient of hundreds of millions of advertising dollars per year.

In terms of public health, what was gained by banning cigarette advertisements on television? Looking back today, the answer may be "nothing." In their desperate attempt to "do good," the public health forces may well have set back their own cause. At least with the anticigarette ads on the air, the country was demonstrating disillusionment with the cigarette. Now few stations would offer free air time to the American Cancer Society or others. But the tobacco advertisers saturated magazines, newspapers, posters for taxicabs and buses and other such outlets.

Ignore and suppress!

Surely tobacco companies were confident that once they dumped millions of their dollars, previously committed to television, into the print media, they would gain the benefit of silence, or at least subdued coverage of the smoking and health issue. And they were right.

Two surveys, one in 1978 by a reporter for *Columbia Journalism Review* and another in 1982 by the American Council on Science and Health, concluded that magazines were indeed influenced by the advertising revenue they receive. While there is no direct evidence that the tobacco companies actually tell editors not to write about tobacco, it is clear that the don't-bite-the-hand-that-feeds-you mentality is operating.

Time to diversify!

The tobacco flagship was experiencing some leaks, and just in case it sank, the cigarette industry needed a safe harbor. By 1969, Philip Morris owned the American Safety Razor Co., Burma-Vita Co., a maker of shaving cream, and Clark Brothers Chewing Gum. In 1968 it went into the candy business by becoming the U.S. distributor for Rowntree and Co., an English firm. Lorillard Corp. (Old Gold) by this time owned part of Loew's Theatres, Inc., and derived $600 million annually from non-tobacco products. R.J. Reynolds bought Hawaiian Punch and other related drinks in 1963, and two years later became the proud owner of Brer Rabbit and Vermont Maid Syrups, and My-T-Fine Desserts. Then they went into the Chinese food business (Chun King), Latin Foods (Patio Foods) and others—including petroleum and packaging. American, which made millions with Old George Washington Hill's Luckies, took over James Beam Distilling Company, Duffy-Mott and Bell Brand Foods. Indeed, in February, 1969 they changed their name to American Brands, Inc. The turn toward diversification continued during the next decade. (See Appendix A for more information on tobacco company diversification.)

Some critical questions were unanswered. Should an industry be at liberty to promote a product that 70 million U.S. smokers want, even if it endangers their lives? Do cigarette makers have any responsibility to the public? And should the federal government restrict the industry in any way? In 1960, Dr. Max Finkelstein predicted with an appropriate sense of alarm:

> Hundreds of thousands of men will die of primary lung cancer within the next decade... men who would not have developed this disease at all had they not smoked cigarettes. Unless the carcinogen is removed, or a cure for cancer is found, the toll eventually will be fantastic and catastrophic.

It was. And it still is—not only for men but for women, too!

12

Joe Califano and the Politics of Smoking in the '70s

Once a 3-pack-a-day smoker, Mr. Califano kicked the habit after his 11-year-old son asked him to stop as a birthday present.

With some notable exceptions, the 1970s whispered the sounds of silence in cigarette land. Gone were the repetitive tunes, and the sailors, rowers, drivers, swimmers, tennis players and others who had been earning a "healthy" living doing television and radio ads. Whereas Sam Spade had smoked through every scene, Kojak, the rough cop, was sucking lollipops instead.

The '70s and the cigarette can be summarized fairly quickly. They were a time of even more revelations about the weed. The time bomb effect of smoking became apparent on women who—as advertised—had indeed "come a long way" with dramatically rising rates of lung cancer, heart disease and emphysema. Female smokers were jolted by the discovery that much of the risk of heart disease and stroke, previously attributed to oral contraceptives, could be laid squarely next to the ashtray.

The '70s produced a burst of interest in nonsmokers' rights and the slogan, "Your cigarette smoking may be hazardous to *my* health." A study published in the *New England Journal of Medicine* found that nonsmokers regularly exposed to smoke showed a small but significant degree of respiratory impairment, similar to that of people who smoked half a pack daily. As "No Smoking" sections became more popular, conflicts occasionally broke out between persistent smokers and others who wanted to breathe clean air. In one case, described in Chapter 19, a disturbance on an airplane became so intense that the pilot made an unscheduled landing to resolve it.

The '70s produced tremendous concern about the increasing rates of cigarette smoking in young people. Robert S. Morrison, M.D., a professor at the Massachusetts Institute of Technology and Cornell University, told *Intellect* in 1977:

The most important public health problem today is America's failure to communicate the cancer and heart disease risks of cigarettes effectively enough to make teenagers decide not to start smoking. Studies generally show that about half of high school boys, and about a third of high school girls became even more attached to the habit.

During the '70s there was relatively little legislative or regulatory activity related to cigarettes. The one exception was the FTC's success in ordering the six major manufacturers of cigarettes to carry a line in all print advertising which read "Warning: The Surgeon General Has Determined That Cigarette Smoking is Dangerous to Your Health." The warning had to be printed black on white in a type size scaled to the size of the ad.

Califano and controversy

If those in the pro and con camps of the cigarette issue were in the business of giving awards, HEW Secretary Joseph A. Califano would be a shoo-in for the Most Colorful Performer of the Decade. Once a 3-pack-a-day smoker, he kicked the habit on October 21, 1975, after his 11-year-old son asked him to stop as a birthday present.

When Mr. Califano was designated Secretary, President-elect Carter told him that he wanted to make a major impact on public health. Armed with his personal interest in smoking, having been through the rigors of giving it up, Califano dug into the facts. He was informed by researchers at HEW that smoking was a factor in some 320,000 deaths each year, including 220,000 from heart disease, 78,000 from lung cancer and 22,000 from other forms of cancer. He was also told that smoking added $5 to $7 billion to health care costs and between $12 and $18 billion in lost productivity, wages and absenteeism. To prevent disease in this country, what better way would there be than to attack cigarette smoking?

Califano assembled an advisory task force and set a press conference for January 11, 1978, to announce his plans. There was much speculation in the press that he would come down hard on cigarettes. Suddenly, the President's Special Assistant on Health Issues, Dr. Peter Bourne, began to sound like he was reading press releases prepared by the Tobacco Institute.

In November 1977, Bourne had stunned the American Cancer Society by attacking programs which he felt made "outcasts" of people who smoked. Addressing the ACS Ad Hoc Committee on Tobacco and Smoking Research, Bourne said, "If our behavioral research shows that a high percentage of cigarette smokers began the habit in a rejection of authority, then we must be sure that the imposition of government authority will not do more to increase their dependence rather than encourage them to quit"—another quote suitable for the *Tobacco Apologist's Guide to Chop-Logic*.

In his extremely candid book, *Governing America: An Insider's Report from the White House and the Cabinet,* Califano tells all about the White House pressure to silence him. He reveals that Dr. Bourne made a number of efforts to break through the "no leaks" security to get an advance copy of the Califano anti-smoking plan. Califano wrote: "Suspicious of his motives, we refused to give it to him. He called me the evening before the speech, urging me not to mount a major anti-smoking campaign."

The tobacco folks wanted the report too; and when they couldn't get it, they launched a new tactic. They held a press conference to deny and condemn the report—which had not yet been released! On January 11, 1978, Secretary Califano did reveal his plan—citing the fact that cigarette smoking kills more than 320,000 Americans each year—and announcing a $29.8 million HEW program centered in the newly created Office on Smoking and Health. Two-thirds of the money would support research into the reasons why people start smoking, the factors that enable some to quit easily while others are unable to do so, and the efficiency of various "withdrawal" programs. The remaining funds would go for public information programs. Noting that nonsmoking is a cornerstone of preventive public health, Califano later shared his ideas with Senator Edward Kennedy's health subcommittee, which was considering two bills relating to smoking.

During his press conference, Califano said that his greatest concern was that "so many Americans start smoking at a very early age, as a result, in part, of expensive cigarette campaigns . . . (which) portray smoking as attractive and mature." To call attention to the problem's seriousness, Califano's speechwriters worked some ear-catching rhetoric into the talk, calling smoking "slow-motion suicide" and designating it "Public Health Enemy Number One."

Califano's plan was immediately attacked by a number of consumer advocates as well as by tobacco representatives. Dr. Sidney Wolfe, director of the Nader-inspired Health Research Group, claimed that the budget was too small to change people's smoking habits since tobacco companies spend almost 20 times as much to promote their products. John Banzhaf, III, Executive Director of Action on Smoking and Health, complained that Califano had "labored mightily and brought forth a mouse," and that the country really needed a proposal that was "gigantic, horrendous, hardhitting." Clara Gouin, leader of the Group to Alleviate Smoking Pollution (GASP), said at a press conference that HEW was sending "a mosquito against a fleet of jets."

Tobacco's friends took an opposite tack. Senator Jesse Helms (R-NC) and Representative L.H. Fountain (D-NC) criticized Califano's stand as unreasonable. "At a time when farmers and all the consuming public are threatened with rapidly rising costs, including a large Social Security tax increase," said Helms, "here comes Secretary Califano demonstrating a callous disregard for economic realities, and particularly for the economy of North Carolina." Noted James Bowling, senior vice-president at Philip Morris, Inc., "I really believe the Califano type of thing is a tragic disservice to science. The real need is to find answers." Quipped Tobacco Institute Vice President, Bill Dwyer, "America, beware if Joe Califano ever decides to give up drinking or other pleasurable pursuits."

The political fallout from Califano's anti-smoking effort was intense. The tobacco industry financed bumper stickers announcing "Califano is Dangerous to My Health" and papered highway billboards all over the South with "Califano Blows Smoke." The North Carolina Department of Agriculture called it a "misguided crusade." South Carolina's Commissioner of Agriculture said farmers were complaining that their tax dollars were being used "to ultimately destroy their means of livelihood." Virginia's Department of Agriculture rounded things out with a telegram to Califano, expressing outrage about "the unwarranted attack... supported at best by vague scientific evidence."

Califano's hands were slapped by an anonymous White House spokesperson who felt the Secretary had not thought through "the political details" or obtained "political clearance" before acting. After the President met with Governor Jim Hunt and Senator Robert Morgan of North Carolina, word came indirectly to Califano to "cool the rhetoric on the anti-smoking campaign, to stop using phrases like 'slow-motion suicide,' and to stop speaking about the subject of smoking." Hunt and Morgan had said that it would be preferable for Califano to "go after alcohol." The President himself never mentioned to Califano the meeting with Hunt and Morgan.

Despite President Carter's long history of interest in preventive medicine, he had always avoided public discussion of tobacco's health hazards. While Governor of Georgia, he had published a pamphlet on "killers and cripplers" which discussed the causes of heart disease, cancer, stroke and other ailments without once even mentioning cigarettes or tobacco.

The Califano-Carter controversy began heating up in August, 1978, when Mr. Carter took a sentimental journey through cigarette land. He waxed eloquent in describing his tobacco-farming ancestors (but did not mention that his own father, an avid cigarette smoker, had died from lung cancer). He talked of "backbreaking labor" and "honest work." He mentioned God and all the churchgoing families, and finally said that there was no incompatibility between promoting good health and promoting a good tobacco crop. He noted that "Joe Califano did encourage me to come here. He said it was time for the White House staff to start smoking something regular," a reference to the reports that marijuana use was rampant among Carter's staff. Then, for some reason, he promised that his administration would "make the smoking of tobacco even more safe than it is today."

After the President's speech the Tobacco Institute declared, "We could not have written it better than that."

Yes, the tobacco people could not have had better representation than they did with Carter and Bourne in the White House. Joe Califano must have been an enormous embarrassment to them. At first they tried to handle him by buffering his comments when talking to the press. "It's not his responsibility to tell a particular American citizen whether he can or cannot smoke," Carter declared. Bourne consistently downplayed the adverse health effects of cigarettes, stating at one point, "No matter how much we may favor prohibition of tobacco products, we are 300 years too late." On another occasion, while speaking of education programs to combat smoking, he claimed that "such efforts are doomed to failure." Bourne, who was associated with Carter while the latter was Governor of Georgia, actually speculated about "benefits" from smoking. "There should be no automatic assumptions in such research that there are no beneficial effects of tobacco use. It may be that certain of the chemical breakdown products of tobacco have beneficial or mixed effects."

A few days after Carter's speech in tobacco country, front-page newspaper headlines announced the results of a 14-year study of smoking and health financed, ironically, by the tobacco companies. "Cigarettes Are A Major Health Hazard: Smoking Causes Lung Cancer, Heart Diseases, May Be Tied to Ulcers."

National Cancer Institute researcher, Dr. Gio Batta Gori, then announced a short-term study showing that some of the new low-tar cigarettes could be smoked in "tolerable" numbers without apparent bad effects on smokers. However, the media distorted what he was trying to say. Gori's point was that toxic substances in certain brands of cigarettes had been reduced to a degree that smoking just a few of them—although still harmful to health—might not have a detectable effect on the published death rates. The press generally missed the fact that he added, "The only safe cigarette is an unlit cigarette."

Coupled with the President's comments in North Carolina, the Gori press conference drew considerable coverage, and prompted the National Cancer Institute Director, Dr. Arthur C. Upton, and National Heart, Lung and Blood Institute Director, Dr. Robert I. Levy, to denounce any inference that scientists now believe that low-tar cigarettes may be considered "tolerable" or "safe" to use.

Another blast from the Surgeon General

On January 11, 1979, Surgeon General Julius Richmond released a 3-inch-thick book calling the case against cigarette smoking "overwhelming." (Again, the tobacco industry held a press conference the day before to call the document, which they had not yet seen, "more rehash than research.") Califano teamed up with Dr. Richmond in the report's release and conducted a media blitz on cigarettes and health—and as he reports, "The impact was stunning." A survey revealed that during the next two weeks, more Americans tried to quit smoking than had during any other 2-week period since the release of the Surgeon General's first anti-smoking report in 1964.

That was when the President's "even safer" comment backfired. During the media coverage of the new Surgeon General's report, the networks showed footage of Carter smiling at the tobacco leaves in Wilson, North Carolina, talking about making smoking "even safer." After the broadcasts, Califano reports, Vice President Mondale called and said, "Jeez, those guys in the White House really have it positioned—the President is for cancer, and you're for health."

House Speaker Tip O'Neill told Califano in late 1978, "You're driving the tobacco people crazy. These guys are vicious. They're out to destroy you." By the spring of 1979, Secretary Califano knew that his days in the President's Cabinet were numbered.

As Fall approached, there was considerable concern about the effect of the anti-smoking programs on Carter's chance for re-election. Raymond J. Mulligan, president of the Liggett Group (who characterized the President as "a silly ass") predicted, "The President couldn't get himself elected to a sewer commission in North Carolina because of Joe Califano." In April, 1979, Senator Edward Kennedy told Califano, "You've got to get out of the Cabinet before the election. The President can't run in North Carolina with you at HEW. He's going to have to get rid of you." And three months later, the President did so.

Tobacco roars on

By the mid-1970s, few people believed what the tobacco industry was saying in its defense, probably including the spokesmen themselves. From the way they spoke—generally in one-liners and "cute" asides, it was apparent that most of the spokesmen actually rehearsed and memorized their lines and answers to potentially embarrassing questions. This is evident in any magazine or newspaper story in which a tobacco industry executive was interviewed. The words are almost always *exactly* the same. There is no spontaneity, lest inadvertently one of them drift from the party line.

What it lacks in facts, however, it makes up for with money. Since the mid-'70s, the industry has conducted a well-financed campaign to discredit the opponents of second-hand smoke and to cast doubts on the scientific evidence linking smoking and disease. The campaign has included "educational" ads in major magazines, carrying the headline, "A Word to Smokers." "Freedom of Choice is the Best Choice," one ad suggests, "It is no secret that there are some folks these days who are trying to build walls between smokers and nonsmokers." The message would have us all live together happily—somehow—as though the problems of lung cancer, heart disease and annoying cigarette smoke can be dissipated by a spirit of mature togetherness.

William Dwyer, Assistant to the President of the Tobacco Institute, suggested to broadcasters and editors during a nationwide tour that there was "another side" to the smoking and health story. Soon he was joined by a Connie Drath. Between 1974 and 1976, Dwyer and Drath made more than 500 single or joint appearances, including appearances on major TV network programs, where they expressed concern about the rights of smokers who were prohibited by law to smoke in public places, about the costs of such laws, and about the American tradition of freedom of choice. They denied even the possibility of a health hazard from smoking, charging:

> A widespread anti-tobacco industry is out to harass 60 million Americans who smoke and to prohibit the manufacture and use of tobacco products... Outrageous and medically unsubstantiated assertions made by well-financed and highly-organized groups opposed to smoking are disputed by many men and women of science.

Why do you suppose that television stations, particularly network stations, allow themselves to be manipulated in this manner? Major American industries customarily conduct media tours to promote such things as soda, soup and shampoo. But these products are safe. Would the networks welcome a guest who declared that the tubercle bacillus has nothing to do with tuberculosis or that people should cough in each other's faces? If you offered enough money, you could probably get a few doctors to say that. But would they be worthy of air time? What about members of the Flat Earth Society? Would the networks welcome *them* with open arms if they decided to go on tour? Isn't it about time for the "tobacco company view" to be given less free exposure?

You gotta have clout

Tobacco men seem willing to do just about anything to protect themselves. Mr. Califano tells the story about the time, at his invitation, Pennsylvania Democratic Congressman Fred Rooney's wife, Evie (who had been a "classmate" of his at SmokEnders) sat on the stage with him during one of his speeches. "When a Tobacco Institute lobbyist saw her, he told her husband that he would never get another dollar from the industry." When Governor Hunt of North Carolina suggested that Califano visit his state and meet with some farmers, North Carolina Congressman Charlie Rose objected, "We're going to have to educate Mr. Califano with a two-by-four, not a trip."

Tobacco men worked very fast in England in 1976 when director Martin Smith and Reporter Peter Taylor released a film called *Death in the West* on a show roughly equivalent to this country's "60 Minutes." Adam Hochschild of *Mother Jones* magazine, one of the first Americans to see it, gave this description:

> [The] searing half-hour film simply intercuts three kinds of footage. The first is old Marlboro commercials—cowboys lighting up around the chuckwagon . . . and so forth. The second is interviews with two Philip Morris [the company that makes Marlboro cigarettes] executives who claim that nobody knows if cigarettes cause cancer. The third is interviews with six real cowboys . . . who have lung cancer or, in one case, emphysema. And after each cowboy, the film shows the victim's doctor testifying that he believes his patient's condition was caused by . . . cigarette smoking.

Irving Rimer, Vice President for Public Information of the American Cancer Society, has also seen the film. He comments:

> *Death in the West* told the truth about cigarette smoking in graphic terms never before shown. In one simple direct attack, it debunked all the glamour from cigarette advertising. Our staff viewed it and, to put it mildly, were in unanimous agreement that it should be shown over television in this country. If Americans were given the opportunity to view it, the impact on the sales of cigarettes would be devastating.

Several American groups expressed interest in showing *Death in the West*, including the American Cancer Society, which wanted to use it in their anti-smoking programs, and CBS television, which wanted to incorporate parts of the film into a news segment for *60 Minutes*. Unfortunately, their plans were quickly blocked. With the obvious intention of preventing the film from being shown to enormous American audiences, Philip Morris filed suit in a British court against Thames Television, producers of the film. The suit claimed that the company had been deceived and had allowed Marlboro commercials to be used in the film without knowing that cigarettes would be depicted unfavorably.

Mother Jones, reporting this story in January 1979, thought this "a little hard to believe, given the fact that filmmaker Peter Taylor had previously made several . . . films about cigarettes . . . which were, to put it mildly, not pro-industry." But the court issued an order preventing the film from being shown or even discussed by Taylor.

Hochschild also reported that Philip Morris seemed embarrassed by its executives' on-camera attempts to defend tobacco medically. This is curious because company spokesmen frequently make similar statements. For example, the Philip Morris' own 1979 Annual Report claimed that "No conclusive clinical or medical proof of any cause-and-effect relationship between cigarette smoking and disease has yet been discovered."

The film might have been freed had the lawsuit come to trial, but the cost of defending the film would have been prohibitive. The case was eventually settled out of court, with all copies of the film given to Philip Morris as part of the settlement. Peter Taylor commented that "*Death in the West* will never live to ride again." Neither will four of its six cowboys who have since died of their illnesses.

Fortunately, *Death in the West* was reborn in the United States and aired twice on KRON-TV in San Francisco during 1981. Audience response was overwhelmingly favorable, according to Dr. Stanton Glantz, President of Californians for Nonsmokers' Rights, who was responsible for providing a mysteriously remaining copy of the film to the station. Widespread distribution of the film apparently began in the Fall of 1983, but despite more than 50 showings—mostly on Public Broadcasting Stations—Philip Morris has been silent. Obviously, the company realizes that in America, attempting to suppress such a film will draw enormous publicity and encourage people to watch it. In December 1983, Pyramid Film and Video of Santa Monica, California, began widespread marketing of films and videotapes of *Death in the West* with net proceeds earmarked for the California Nonsmokers' Rights Foundation.

Peter Taylor is a hard man to keep down. His more recent film about tobacco, *A Dying Industry*, was shown on British television in April 1980. This, too, pulls no punches. It quotes the British Prime Minister saying, "My health ministers and I are in no doubt that smoking is the major preventable cause of illness and premature death in the U.K." The film shows a patient having his leg amputated as a result of smoking-related peripheral vascular disease. It describes a European tobacco industry committee which claims that smoking has not been shown to cause human disease, then interviews a prominent scientist, formerly employed by a tobacco firm, who says that scientists within the tobacco industry "refer to that particular organization as the Flat Earth Society." However, this film does not focus mainly on the victims of cigarettes but on the manufacturers and the tactics they use in their fight for economic survival.

In *A Dying Industry*, the interviewer violated the tobacco industry ground rules—to the effect that only "precleared" questions will be asked—by asking a tobacco executive a direct and embarrassing question. The executive panics on camera and looks toward his colleagues for assistance. The film ends there.

The tobacco industry (as it frequently reminds the public) funds a great deal of research. At one time, it supported a famous institute in Germany which was studying the possible relationship between cigarette smoke and atherosclerosis in pigs (which have a respiratory system very similar to that of humans). In June 1976, the institute suddenly shut down, supposedly because of a sudden withdrawal of funds for further research and illness of the institute's director. However, according to a scientist interviewed anonymously in the film, funds were plentiful and the director enjoyed the best of health. The scientist and most of his colleagues believed that the shutdown was prompted by results emerging from the pig research which implicated smoking in causing atherosclerosis in this critically important species.

The tobacco industry's consistent position that the relationship between smoking and health is still "controversial" is obviously a survival tactic. *A Dying Industry* suggests that the reason for this position is that acknowledging the danger might subject the industry to a ruinous avalanche of lawsuits by injured cigarette smokers and their survivors. Of course, it would also cut into sales.

In the film, a tobacco industry spokesman says that "The social acceptability issue will be the center battleground on which our case will be lost (in the U.S.)." Taylor then describes some industry efforts to fight attempts to portray smoking as a dirty, unattractive habit, including impeding the efforts of GASP, ASH, and other advocates of nonsmokers' rights to have laws passed restricting smoking in public places. The film's transcript also notes that when the U.S. Department of Health, Education and Welfare decided to follow the example of many major businesses and subsidize its employees' participation in SmokEnders, the tobacco lobby and its Congressional contacts quickly pressured the department into discontinuing the program.

A Dying Industry also claims that an early effort by a major U.S. insurance company to launch a life insurance plan with discount rates for nonsmokers was cancelled before the first commercial hit the airwaves. The film claims that tobacco growers in North Carolina accused the company of discrimination, and state insurance officials began an investigation—all before the presumably secret new campaign had been made public. How could the tobacco lobby have known about the insurance plan? The film provides proof to suggest that there was a leak in the advertising agency which handled both the insurance firm and a major tobacco company.

Dr. David Fletcher, a California public health authority, was asked by an editor of the *Journal of the American Medical Association* to write about *Death in the West* and its health implications. In a pattern consistent with the AMA's ambivalence about speaking out on the dangers of tobacco, the piece was "spiked" on the grounds that it would render the AMA "vulnerable to legal action."

Hear no evil, see no evil; but it's OK to read about it

Back in the good old days before the Public Health Cigarette Smoking Act of 1970 when into effect, a very healthy round of anti-smoking advertising portrayed tobacco as a smelly, addictive, dangerous substance. These ads were public health education at its best. So were the attempts during that period to give cigarettes the ridicule they deserve. One showed people literally coughing their heads off. Another was Bob Newhart's imaginary telephone conversation in which Sir Walter Raleigh tries to persuade his British superiors who had never seen a cigarette that they would sell well in England. It went something like this:

> You stick it where, Walt? In your mouth? and you mean you roll this cylinder, light it and then breathe the smoke? Is that right, Walt? Say, Walt, baby, I'll tell you how you can accomplish the same thing. Light a pile of leaves, stick your head in the smoke and breathe in it.

Those days are gone. Public interest was certainly not served by the action that removed the anti-smoking ads. How did the tobacco companies fare? Did they continue to thrive without TV and radio to help them out? They certainly did!

The industry used a number of strategies. First, using health concerns of smokers to their advantage, manufacturers greatly expanded their product lines to include a wide variety of low-tar, low-nicotine cigarettes. Their promotional campaigns revolved increasingly around the low-tar and nicotine content of their products. By the end of the decade, almost all brands offered a "light" alternative. Second, ignoring warnings from the FTC, they increased the use of advertising themes that denied the health hazard issue. Male smokers were portrayed as virile. "Camel Country," we were told, is "where a man belongs," whatever that means. Female smokers were portrayed as beautiful, sexy, carefree, "liberated" and supposedly enjoying good health. During this decade, more older smokers were shown in the ads to imply, "See, we have been smoking for years and we feel fine." Third, the cigarette industry increased advertising wherever it was legally allowed.

When the congressional ban went into effect, the industry was spending 205 million dollars annually on television advertising, about 12½ million on radio, and 64 million dollars in magazines and newspapers. Newspaper ad revenues soared from $14 million in 1970 to $241 million in 1979, while magazines rose from $50 million to $257 million per year during the same period. In addition, cigarette companies increased advertising enormously on outdoor billboards, in transit systems, at the point of sale in stores, and in "special events" like ballgames which just happened to display ads on the backfield fences.

Some skeptical Congressmen had predicted that the shift to print would happen in a big way. But the tobacco men denied it. Joseph Cullman of Philip Morris stressed repeatedly during an appearance on CBS' *Face the Nation* on January 13, 1971, that he and his colleagues intended to adhere to the spirit as well as the letter of the law. He even said that a significant portion of the money previously used for TV and radio ads would be spent on "research . . . something in the area of a half to a third of the advertising budget." Back in 1969, Senator Frank E. Moss, Chairman of the Consumer Subcommittee, had asked Cullman what the industry intended to do about advertising once its commercials went off the air. He replied, "I think that is too important a matter for me to answer directly, other than to say that it is a large amount of money and that we will approach it constructively."

Why did the magazines and newspapers accept this increasing number of ads? Do you think the bottom line was greed? Writing in the *Washington Monthly*, James Fallows suggested that selling advertising space to cigarette companies is no less demeaning than selling it to "a gang rape club or a Mafia recruiter." Noting that *The New York Times, Washington Post, Time* and *Newsweek* had gotten along without such ads, he said:

> For most of them, taking cigarette ads is not a matter of making profit, but only of maximizing it. The extra five or ten percent this adds to their revenues comes to an enormous cost in hypocrisy and, let us say it outright, contempt for human life. Everything else they say will be cheapened until they stop.

What do the magazines say for themselves? *Business and Society Review* asked a few editors how they justified running ads in light of the grim medical evidence?

Ralph P. Davidson, publisher of *Time*, replied, "Our decision to accept this advertising is based on the fact that cigarettes are lawful items of commerce."

Richard A. Shortway, publisher of *Vogue*, noted that he was aware of the hazards and added, "I feel confident that there is today not one reader of our magazine who is unaware of the inherent dangers of smoking; but I believe that if this reader wants to smoke, it is up to him or her and not us."

David M. O'Brasky, former editor of *Esquire*, responded:

> Certain products could fall into the classification of being detrimental to one's health if taken in excess—tobacco and alcohol would fit into that designation. Or we might be offered advertising for countries whose government policies do not please us. In these situations, it is our view that the reader of *Esquire* must make his (or her) own considered judgement. Therefore we do indeed accept tobacco and alcohol advertising.

Arthur M. Hettich, editor of *Family Circle*, said, "The tobacco industry is supported by our government and there is no official prohibition on smoking."

No one admitted taking the ads exclusively for the money.

By the end of the decade, a record-breaking *one billion dollars* was being spent on cigarette advertising—not exactly what tobacco companies promised when the ban on broadcast advertising was being considered.

What was the impact of this switching of advertising contracts? R.D. Smith, then managing editor of the *Columbia Journalism Review*, summarized it in his 1978 article, "The Magazines' Smoking Habit":

> A survey of the leading national magazines that might have been expected to report on the subject, reveals a striking and disturbing pattern. In magazines that accept cigarette advertising, I was unable to find a single article in seven years of publication, that would have given any clear notion of the nature and extent of the medical and social havoc being wreaked by the cigarette smoking habit. The records of magazines that refuse cigarette ads, or do not accept advertising at all, were considerably better.

For example, *Newsweek*, in January 1976, ran a cover story entitled "What Causes Human Cancer." Their list of known or suspected causes reads as follows: 1) food and drink, 2) drugs, 3) radiation, 4) the workplace. They gave extensive coverage to each of these causes, but gave no section to cigarettes, even though the story included figures showing that lung cancer accounted for the largest proportion of all cancer in males. *Esquire's* love affair with the cigarette included an article in 1975 by Richard Selzer, M.D., entitled "In Praise of Smoking."

In 1978, the U.S. tobacco industry sold 616 billion cigarettes to the nation's estimated 50 million smokers, a 0.6 percent increase from the previous year. A *Business Week* article predicted in 1979 that "Cigarette sales in the U.S. seem bound to begin a long slow decline."

That prediction may have been premature.

13

Is There a Safe Tobacco Product?

The mortality of smokers and nonsmokers is exactly the same: 100 percent. The difference is in the timing.

Are filter or low-tar cigarettes safe? The tobacco folks suggest that the answer is "yes." Indeed some of their current ads sound almost like public health announcements. But they are not!

Most smokers are probably confused about this issue. A 1980 Roper study asked people to judge the following statement: "It has been proven that smoking low-tar, low-nicotine cigarettes does not significantly increase a person's risk of disease over that of a nonsmoker." Although the statement is *false*, 36 percent of smokers thought it was true, and an additional 32 percent said they didn't know.

Cigarette smoke is composed essentially of tar, nicotine, moisture, air, carbon monoxide and other gases. The FTC measures the amount of tar and nicotine with a "smoking machine" that draws smoke from the cigarette and traps most of the particles. This material is weighed, and the nicotine and moisture are extracted and measured. The rest of the trapped material—a crude mixture containing an estimated 4,000 different compounds—is the "tar." Because tar promotes cancer, low-tar cigarettes might be expected to be safer. But the issue is not that simple. In his 1981 report, *The Health Consequences of Smoking: The Changing Cigarette*, then Surgeon General Dr. Julius Richmond summed up the facts as follows:

1. There is no safe cigarette and no safe level of consumption.
2. Smoking cigarettes with lower yields of "tar" and nicotine reduces the risk of lung cancer and, to some extent, improves the smoker's chance for longer life, provided there is no compensatory increase in the number smoked. However, the benefits are minimal compared to those of giving up cigarettes entirely.

3. It is not clear what reduction in risk may occur for diseases other than lung cancer. For heart disease (the largest component of excess mortality caused by smoking), low-tar/nicotine cigarettes do not appear to be any safer than their high-tar counterparts. There is not enough information on which to base a judgment in the case of chronic obstructive lung disease. And there is no evidence that changing to a lower "tar" and nicotine cigarette has any effect at all on reducing the danger to the developing baby during pregnancy.

4. Carbon monoxide in cigarette smoke is known to be harmful. However, it is not known how varying the carbon monoxide level affects the risks involved in smoking.

5. Smokers may increase the number of cigarettes they smoke and inhale more deeply when they switch to lower yield cigarettes. Such compensatory behavior may negate any advantage of the lower yield product or even increase the health risk.

6. "Tar" and nicotine yields obtained by present testing methods do not correspond to the dosages that the individual smokers receive; in some cases the tests may seriously underestimate these dosages.

7. A final question remains unresolved: Do the new cigarettes being produced today introduce new risks through their design, filtering mechanisms, tobacco ingredients, or additives? The chief concern is additives (see Chapter 20). The Public Health Service has been unable to assess the relative risks of cigarette additives because manufacturers won't reveal what they are.

In 1982, *The New York Times* noted that Brown and Williamson had complained to the FTC that American Brands, Inc., Philip Morris, U.S.A., and R.J. Reynolds Industries were engaging in deceptive advertising. While promoting very low-tar cigarettes packaged in flip-top boxes, the three were also marketing cigarettes containing 10 to 100 times more tar—in look-alike soft packages. The *Times* also reported that Brown and Williamson's much-publicized low-tar Barclay was designed to fool the FTC's smoking machines. The machines preserve the Barclay filter—but the human lips probably destroy it, giving smokers heavy doses of just what they were trying to avoid. In January 1983, *Consumer Reports* noted that while the Barclay ads claimed "1 mg. of tar," smokers actually got 3 to 7 times as much. More recently, the FTC announced that there is "a significant likelihood" that Kool Ultra and Kool Ultra 100s have a similar problem with their filters.

Adding it all up, using "safer" cigarettes is probably like deciding to jump out of a window on the 36th floor of a building instead of the 39th. Do other forms of tobacco carry the same risk? The answer here is no, that cigarettes are much more dangerous than the others. But all forms of tobacco use increase the risk of disease.

"Smokeless tobacco"

An estimated 22 million people in the U.S. use smokeless tobacco—snuff and chewing tobacco. Moist, smokeless tobacco is the only segment of the tobacco industry that is still growing today.

The U.S. Tobacco Company plans to spend $10 million in 1984 to push its newest product, Skoal Bandits, a snuff product which, like tea, comes in little pouches that eliminate the problem of loose tobacco bits in the mouth. All three major television networks will get commercials, and print advertising will appear in such mass-media circulation publications as *TV Guide, Parade, Family Weekly, National Star, Newsweek,* and *Sports Illustrated.* "Red Man," the country's best-selling brand of chewing tobacco, is now being advertised on television stressing the "macho" image.

Who stars in these new commercials? None other than our nation's sports and music heroes. The December 1983 *New York State Journal of Medicine* reported that:

> The array of celebrities employed to cultivate the puberty rite of tobacco use includes baseball players George Brett, Carlton Fisk, Catfish Hunter, Sparky Lyle, and Bobby Murcer, and football players Terry Bradshaw, Nick Buoniconti (now a tobacco and candy distributor), Earl Campbell, Joe Klecko, and Laurence Taylor. Singer Charlie Daniels appears on a high-priced collection of paraphernalia for a brand of snuff, Skoal.... A race car driver, Harry Gant, drives a car painted with "Skoal Bandit"; his entourage includes a group of cheerleaders called the Skoal Bandettes. In 1983, a commercial record sung by New York Yankee Bobby Murcer, was frequently played on teenage-oriented radio stations.

Other ads urge those who want to stop smoking but don't want to give up tobacco to "take a pouch instead of a puff."

Smokeless tobacco, especially snuff, is becoming the craze of young males across the nation. It apparently creates a masculine and mature self-image among young men and boys on junior high school, high school and college campuses. Even some who don't care for snuff still carry the small round tin in a rear blue jean pocket because once the circular shape of the tin is worn and fades into the blue denim, it supposedly symbolizes virility, machismo, and toughness.

A study cited in a September 1980 issue of the *Journal of the American Dental Association* found that as many as one-third of the members of the varsity football and baseball teams were chewing, dipping, or both. For the past four years the Beech-Nut (chewing) Tobacco Company and the Monroe County Indiana Fair Board have sponsored a "Tobacco Spitting Contest" as part of the annual 4-H County Fair. Until recently, the contest was open to "everyone over the age of 13." But as the October 1983

Smoking and Health Reporter noted, only a handful of contestants were *not* minors.

Chewers can choose from more than 150 brands. They are dated for freshness, like dairy products, and are generally made from the same kinds of tobacco used in cigars. There are four main types of chewing tobacco: loose leaf, fine cut, plug and twist.

Loose leaf, the most popular of the chewing products, is made almost entirely from cigar-leaf tobacco. It is sold in small packages and classified as either sweet (and heavily flavored) or plain. Fine cut is similar but much more finely cut. Some of it almost resembles snuff, which is finely ground or powdered cigar-type tobacco. Plug is leaf tobacco that has been pressed into flat cakes that look somewhat like brownies. They are either moderately or heavily sweetened by such flavorings as molasses, licorice, honey or maple sugar. Twist is made from stemmed leaves, twisted into small rolls and folded. It, too, is usually flavored. The "chew," in whatever form, is held between the cheek and the lower jar. Chewing causes nicotine to be absorbed into the blood stream and give the user a lift. Chewers can, of course, become addicted to nicotine.

Snuff is usually classifield as either dry or moist, both types of which may be sweetened, flavored, salted and/or scented. Popular flavorings include spearmint and wintergreen. Moist snuff is somewhat coarser than dry and is by far the more popular. In times past, snuff was snuffed (sniffed) through the nose. Now it is generally "dipped" by tucking a pinch into the mouth between the gum and lower lip. Snuff is absorbed so rapidly that tobacco chewers sometimes complain that it gives them too much of a jolt. "It's like mainlining nicotine," noted one baseball player.

Of course, the chew and its juices must be disposed of periodically, which is not the most aesthetic of procedures. The reason for spitting rather than swallowing is that tobacco juice is not considered tasty, even by tobacco freaks. In fact, it has been compared to battery acid and worse. To paraphrase D. Keith Mano writing in *National Review* a few years ago: Swallow a little and you don't want to eat; swallow a lot and you don't want to live. Every chewer seems to have a favorite story about the time he, or someone he knew, accidentally swallowed an entire quid. Enough said.

The use of smokeless tobacco is not a passing fad. Overall sales have increased approximately 11 percent per year since 1974. During 1978, nearly 117 million pounds were sold in the United States—92.3 million pounds of chewing tobacco and 24.3 million pounds of snuff. U.S. Tobacco, the country's largest snuff manufacturer (Skoal, Happy Days and Copenhagen), sold 1¼ million cans of moist smokeless tobacco every business day of 1979. That's an increase of 95 percent since 1969, mostly reflecting new users in the 18-30 age group. In 1977, overall sales of loose-leaf tobacco, which accounts for about half of the entire smokeless market, amounted to $325 million wholesale. Retail snuff sales are now about $500 million a year.

Explaining the reasons behind the popularity of smokeless tobacco, David Weiss, a marketing manager for General Cigar and Tobacco Company (the manufacturer of Mail Pouch), told *The New York Times*: "safety and convenience." Smokers turn to smokeless tobacco hoping that it is a safe replacement for cigarettes and cigars. Snuff can be chewed or dipped during activities that require the use of both hands. Sailors traditionally chewed tobacco because of the difficulty of keeping their pipes lighted in the wind. And workers in mines and mills find that the moisture helps relieve their dusty mouths and throats.

As with cigarettes, most medical professionals consider the case against snuff and chewing tobacco closed. Although early studies displayed conflicting results, recent ones have linked smokeless tobacco with cancers of the mouth and throat as well as gum disease, tooth decay, and excessive tooth wear. It is not uncommon for leukoplakia—whitish or grayish patches of cells—to develop at the places where the quid is held. This condition is precancerous with a malignant transformation rate between 3 and 5 percent, but it usually goes away if tobacco use is discontinued. The patches of irritation can also become thicker, harder and painful.

Only a few years ago, the first organic carcinogen was isolated from unburned tobacco. Known as nitrosonornicotine (NNN), it is a type of nitrosamine that is highly concentrated in chewing tobacco. NNN and possibly other nitrosamines are formed during the curing process because of the heavy concentration of naturally occurring nitrate in tobacco.

Several studies have linked smokeless tobacco with cancer of the mouth, particularly where the chew or snuff was held. "Snuff dipper's cancer" is a term used to describe the oral cancer among women in parts of the rural South where snuff is widely used. Many of these women, employed in textile or apparel factories where smoking was discouraged, turned to snuff. In one study, 90 percent of the women suffering from squamous cell cancer of the mouth were habitual users of snuff. In another, female snuff dippers were found to have four times the risk of oral cancer compared to nontobacco users.

A German study of 15,500 snuff users over a 20-year period disclosed only 2 cases of oral cancer and 12 of severe leukoplakia. Yet in Sweden, a country where the use of smokeless tobacco is increasing rapidly among children and adults, the Cancer Registry of the Board of Health and Welfare reported 33 cases of oral cancer over a 9-year period that were "clearly related" to the use of snuff.

One of the larger studies in the United States was reported several years ago by Dr. James F. Smith of the University of Tennessee's College of Medicine, Department of Otolaryngology. After following more than 15,000 snuff users and tobacco chewers for 10 years, he found not a single case of cancer or serious mucosal change in the mouth. Dr. Smith, who has also performed extensive tobacco research with animals, was led to conclude:

> I believe that the type of snuff used in this country cannot logically be considered carcinogenic in view of the large number of patients who have used snuff for many years with no clinical or histological evidence of tissue change.

However, a 1983 study by researchers at the University of Colorado School of Dentistry looked for changes in gum tissue and teeth associated with the use of smokeless tobacco by teenagers. In a random sample of 1,119 high school students, over 10 percent were smokeless tobacco users. Among the students, the researchers identified cases of hyperkeratosis, gum inflammation, and erosion of the teeth.

Perhaps more research on chewing tobacco has been conducted in India than in any other country, primarily because oral cancer is a major health problem there, with the highest known incidence in the world (21 cases per 100,000 population). Nearly all of the Indian studies show a positive correlation between oral cancer and tobacco chewing; but it is difficult to assess the significance of the findings because Indians usually chew tobacco together with other substances such as betel nuts and lime which can multiply the effect of the tobacco. However, in the tobacco-growing community of Maharashtra, where both men and women do a great deal of chewing, a study has found elevated rates of stillbirths, early deliveries and lower birth weights. The same problems occur in pregnant women smokers in the United States.

More research is needed to determine more precisely the health risks of chewing or using snuff. One of the difficulties in investigating smokeless tobacco lies in distinguishing between "heavy" and "light" users. The amount per chew (or per dip), the length of time it is held in the mouth, and the frequency per day are not easy to sort out and tabulate.

The advertising blitz promoting smokeless tobacco has prompted several groups to urge the Federal Trade Commission to regulate the ads in the same way that cigarette ads are regulated. Doctors Ought to Care (DOC) has petitioned the FTC to require a warning label on all smokeless tobaccos, similar to the one found on cigarette packages. DOC also petitioned the FTC to stop broadcast ads for smokeless tobacco and to require warnings on printed ads. Dr. Tom Houston, DOC's national coordinator stated at the time of the petition, "We take exception to these ads because they are aimed at youth.... We don't like the ads because they get kids to use the product." The Public Citizen Health Research Group has filed a similar petition.

Cigars and pipes

As of 1975, pipe smokers accounted for approximately 12 percent of the smoking population, cigar smokers 20 percent and cigarette smokers 39 percent. It has always been assumed that pipes and cigars pose less of a health risk than do cigarettes, and epidemiological data tend to support these assumptions. Both Hammond and Horn and Doll and associates carried out major epidemiological studies involving thousands of men and found that cigar and pipe smokers had a lower overall death rate than cigarette smokers. However, the overall death rates of cigar and pipe smokers were slightly but definitely higher than those of nonsmokers. Hammond and Horn also found that cigar smokers are more likely to die earlier than pipe smokers. Individuals who smoke *both* cigarettes and pipes or cigars generally have death rates in-between those who smoke cigarettes only and those who smoke pipes or cigars only. (Pipes and cigars *do not* offer a protective effect to cigarette smokers, however. The lower mortality rate among people smoking pipes or cigars in addition to cigarettes reflects the fact that less cigarettes are smoked.)

The discrepancy in death rates between cigar or pipe smokers and cigarette smokers may be partially explained by the different curing and processing methods used to produce the tobaccos. However, a large part of the difference in death and specific disease mortality rates may be due to different practices of inhalation. Cigar and pipe smoke is significantly more alkaline than is cigarette smoke, making it much more of an irritant, and thus, less likely to be inhaled. However, smokers who switch from cigarettes to cigars should be aware that a portion of the average cigar (the first two-thirds) is acidic enough to be comfortably inhaled and that smokers tend to continue their inhalation habits after switching. The same is true for those who smoke both cigarettes and cigars. There is no evidence that when cigar or pipe smoke is inhaled, it is less dangerous than cigarette smoke. In fact, studies have demonstrated that cigar and pipe smoke condensates have a carcinogenic potential equal to that of cigarette smoke condensate when applied to the skin of mice.

Not suprisingly, the most common sites of disease among cigar and pipe smokers are the upper airways and upper digestive tract. In several studies, lip cancer has been associated with the smoking of pipes, but not cigars or cigarettes. Although approximately 1,500 new cases of lip cancer are diagnosed per year, the fatality rate is low because of early detection and surgical accessibility.

Oral cancers, on the other hand, have been linked with all forms of tobacco smoking. Cigarette, pipe and cigar smokers share an equal risk of contracting them. Epidemiological studies have also suggested that alcohol and smoking act synergistically to further increase the risk of developing oral cancer over that of individuals who only drink or only smoke.

Even if a smoker does not inhale, the larynx (voice box) is probably exposed to nearly the same levels of tobacco smoke as is the mouth. While the esophagus may not be directly exposed to tobacco smoke drawn into the mouth, it does have contact with tobacco smoke which is condensed on the membranes of the mouth and then swallowed. Consequently, both laryngeal and esophageal cancer mortality rates are higher in pipe and cigar smokers than in nonsmokers and are approximately equal to those of cigarette smokers.

Pipe and cigar smokers are less likely to develop lung cancer than are cigarette smokers, although, of course, they have higher death rates from this disease than do nonsmokers. Autopsy studies of nonsmokers, cigar or pipe smokers, and cigarette smokers have also indicated that cigar or pipe smoking may be responsible for abnormal lesions in the bronchi of the lung, similar to the pre-cancerous changes caused by cigarette smoking. One Swiss study has even suggested that heavy smoking of some cigars may be responsible for an increased risk of lung cancer similar to that found in cigarette smokers.

While it has been suggested that cigar and/or pipe smoking is related to pancreatic, kidney and bladder cancer, the evidence is not so strong as for other cancers.

Pipe and cigar smokers have only a slightly higher mortality rate from coronary heart disease and strokes than do nonsmokers, but they do experience significantly higher death rates from peptic ulcer. Death rates for both diseases are lower than those found in cigarette smokers.

People who smoke pipes or cigars have more respiratory symptoms such as coughing and wheezing than do nonsmokers, and are more likely to die from chronic bronchitis and emphysema. Again, the risk of contracting and dying from such diseases is less than for cigarette smokers.

In conclusion, although there are definite health risks associated with cigar and pipe use, the hazards are certainly far less than those associated with the use of cigarettes. While it is obviously not wise to use tobacco in any form, the risks can be lessened by switching to smokeless tobacco, cigars or pipes.

14
Reflections on a Burning Issue

Despite overwhelming evidence that its products are deadly, the cigarette industry has never shown any voluntary restraint in promoting them.

No topic more than cigarettes merits the anguished expression of woe that "fate has dealt us a truly cruel blow!" What miserable luck! A form of relaxation and pleasure, thought to be harmless or perhaps even health-promoting when introduced, turned out some 50 years later to be the major cause of preventable illness and death. Yet the disaster of Prohibition makes it clear that popular products cannot simply be banned.

What can America do about the cigarette? For starters, we must face three facts:

1. *It is clear that a substantial number of cigarette smokers are continuing their habit of smoking despite the recognized health risks.* Mark Twain may have spoken for this group when he wrote, "Why, my old boy, when they used to tell me I would shorten my life 10 years by smoking, they little knew the devotee they were wasting their puerile words upon— they little knew how trivial and valueless I would regard a decade that had no smoking in it." The same feeling underlies the Estonian proverb, "It is better to be without a wife for a bit than without tobacco for an hour."

2. *Despite overwhelming evidence that its products are deadly, the cigarette industry has never shown any voluntary restraint in promoting them.* Advertising appears to have three goals: 1) to deny the fact that cigarette smoking is harmful; 2) to stimulate young people to begin smoking; and 3) to make smoking socially acceptable.

3. *Powerful elements in our society have been maintaining a "conspiracy of silence" to avoid dealing with the cigarette problem.* These include publications dependent on advertising and many government officials who are afraid of political repercussions.

Part II of *A Smoking Gun* suggests what we can do to overcome these problems. Let's begin by examining some trends in cigarette consumption.

A slow decline

While the number of cigarettes smoked by Americans reached an all-time high of 640 billion in 1981, the number smoked per person over age 18 has been declining for the past few years. The highest level was 4,286 per capita in 1963. It fell to 3,967 in 1978 and appears to be falling slowly but steadily (see Table 14:1).

Currently about 37 percent of American men and nearly 29 percent of women smoke cigarettes. These represent a significant drop in the rate of smoking among men, from a maximum rate of over 50 percent in the

Table 14:1. Trends in cigarette consumption.*

Year	Number of cigarettes sold (in billions)	Per capita consumption
1900	2.5	54
1910	8.6	151
1920	44.6	665
1930	119.3	1,485
1940	181.9	1,976
1950	369.8	3,552
1960	484.4	4,171
1963	516.5	4,286
1964	505.0	4,143
1965	521.1	4,196
1966	529.9	4,197
1967	535.8	4,175
1968	540.3	3,986
1969	527.9	3,986
1970	534.2	3,969
1971	547.2	3,982
1972	561.7	4,018
1973	584.7	4,112
1974	594.5	4,110
1975	603.2	4,095
1976	609.9	4,068
1977	612.6	4,015
1978	616.0	3,967
1979	621.5	3,861
1980	631.5	3,851
1981	640.0	3,840
1982	634.0	3,745
1983	600.0	3,494

*For U.S. residents and overseas military personnel age 18 and older; based on data from the United States Dept. of Agriculture and the Federal Trade Commission.

early '60s. The rate among women has declined by only a few percentage points, however, from a high of 33 percent in 1965.

Among men, cigarette consumption is inversely related to income and occupational level. Men in professional and technical occupations or who have relatively high incomes are less likely to smoke than are blue-collar workers or those who have relatively low incomes. Some 47 percent of male blue collar workers now smoke compared to 36 percent of male white collar workers.

Educational level seems to be another important factor that influences smoking behavior. The higher the level of education, the lower the likelihood of smoking, although the relationship is stronger for men than for women. According to one survey, more than two-thirds of lawyers, judges, business executives, physicians, dentists, engineers and managers who were once smokers had quit, presumably because they were aware of and understood the tremendous health risks associated with smoking.

Proportionately more blacks than whites smoke cigarettes, but black smokers tend to smoke fewer cigarettes.

Among teenagers, the rate of smoking declined during the 1970s, but recent figures from the National Institute of Drug Abuse indicate a slow-down or perhaps even a reversal of this trend. It is reported that in 1982, 13 percent of male high school seniors smoked half a pack or more of cigarettes per day, while nearly 15 percent of female high school seniors did so. Those who are successful academically tend to smoke less than those who do not perform well.

Mandates for action

The U.S. Department of Agriculture predicts that Americans will still be smoking about 3,100 cigarettes per person in 1990. But despite the general decline in the prevalence of smoking, the cigarette industry is just as profitable and politically powerful as ever. It is clear that the cigarette problem will not just go away by itself. A plan of action and a good deal of patience will be needed to release the tobacco industry's grip on the country. Six specific areas of pursuit merit our attention:

1. *Cigarette smokers should pay their own way.* The free enterprise system ensures that personal freedom should not be violated, and that big government may not prosecute citizens for engaging in activities which they consider pleasurable, legal activity—like gum chewing or cigarette smoking. But that assumes that the cost of the activity is carried exclusively by the person choosing to engage in it. For gum chewing that is true. But for cigarette smoking it is not.

More often than not, cigarette smokers pay the same as nonsmokers for life and health insurance and are treated as equals in workmen's compensation, retirement benefits, social security, Medicare and Medicaid awards. Smokers and nonsmokers pay an equal share of the costs of cigarette-induced fire damage, and the increased cost of goods due to reduced economic productivity of smokers. This diffusion of the economic burden of smoking throughout society is not consistent with the free enterprise philosophy that people should be held responsible for their own behavior.

2. *Cigarette manufacturers should be held legally responsible for the health damage they cause.* Each year thousands of Americans collect damages for illness or injury resulting from exposure to anything from asbestos, cotton dust and industrial chemicals to tampons, toys and other household products. Yet no tobacco company has ever lost a case or paid out one cent in compensation for cigarette-induced illnesses.

3. *The U.S. must face up to the tragedy of cigarette production and sales in the Third World.* While we introspectively ponder the tragedy of cigarettes in America, tobacco peddlers have been exploiting new and profitable markets in the developing world. The situation resembles a movie rerun. Like Americans in the 1920s, Third World residents are being enticed by advertising and sold cigarettes with high nicotine and tar contents but no warning labels. Should we who can foresee the inevitable epidemic of chronic disease 20 to 30 years from now just sit back and watch?

4. *Cigarette advertising must be stopped.* Advertising has always been the main stimulus to cigarette smoking. What advertisers are selling today, however, is not cigarettes *but rather social acceptance of cigarettes.* For America to succeed in reducing smoking behavior, all advertising for tobacco products must stop.

5. *The tobacco industry's smoke screen must be lifted.* And although the industry has formed what is probably the most complex and effective political network in U.S. history, its power can be overcome with determined citizen action. This process will be facilitated if more people who should be speaking out—such as women's groups, religious leaders, consumer advocates and political conservatives— recognize that the problem of smoking deserves their attention.

6. *Nonsmokers' rights must be protected!* Although "second-hand" smoke has less harmful effects than directly inhaled smoke, it is annoying to most people and can cause health problems for some.

Smokers must be given the facts needed to make more intelligent decisions. Although over 90 percent of Americans are aware of the general adverse health consequences of smoking, relatively few people know about the specific cigarette-disease links. Furthermore, there is considerable confusion about the relative safety of various forms of tobacco today—filters versus non-filters, low- versus high-tar and nicotine cigarettes, pipes, cigars—and chewing tobacco. Smokers need the full facts about the risks involved.

Above all, *the smokescreen of rationalizations offered by the tobacco apologists to defend their product should no longer be tolerated!* These irresponsible assertions that the cigarette-disease case is "not proven" and that tobacco is good for America have largely gone unchallenged by scientists and the media. The industry has become more brazen by the day in flaunting its success in surviving the "health scare." Cigarette manufacturers should be made to admit that they are literally getting away with murder and ordered either to stop selling their products or at the very least, turn down the volume of their self-serving rhetoric.

Facing up

It isn't easy to face up to the fact that a commonly used product, one that millions of Americans depend on for income, is at the same time killing people. It is particularly difficult to do so for those who have an economic stake at risk. That is why *The Charlotte Observer* should be congratulated for doing just that: facing up. On Sunday, March 25, 1979, the *Observer* dedicated a 20-page section to cigarettes under the title "Our Tobacco Dilemma: North Carolina's Top Crop: Part of Our Lives, but Bad for Health." The special section began as follows:

> Nourished by North Carolina's peculiar combination of climate and soil, the first green shoots of this year's tobacco crop already tint new plowed seed beds. By summer's end, we will have reaped the harvest—leaves whose cured golden color symbolizes what tobacco and its prime product, cigarettes, have meant to our economy and culture.
>
> But this year, with conclusive new evidence that smoking robs people of their health, the harvest is beginning to represent something else. Tobacco has become our dilemma, a matter of stark contradictions; it is a positive part of our culture and economic lives whose end product—cigarettes—kill some of us.

All of us can learn from this candid statement of the tobacco dilemma. We must also realize that all of us can contribute to its solution. The next seven chapters offer specific recommendations.

15

Smokers Should Carry Their Load

Cigarette-related diseases are responsible for more than $11 billion per year in medical expenses and $36 billion in lost productivity.

A 1978 study by the Roper Organization concluded that the greatest threat to the tobacco industry to date has been the growing public concern that cigarette smoking is dangerous not only to smokers but to nonsmokers as well. The same argument can be applied to the subject of cigarette economics! This chapter describes in simple terms the economic burden of smoking—and how nonsmokers are being unfairly forced to share it.

The cost

According to 1978 estimates, smoking accounts for nearly 8 percent of all direct health care costs and over 11 percent of the total direct and indirect cost of disease in the United States. Direct costs include those incurred in prevention, detection and treatment of illnesses caused by smoking. They also include costs of rehabilitation, research, training, and capital investment in medical facilities. Indirect costs include earnings lost through disease and death, which represent losses to the gross national product (GNP).

Health economists Stuart O. Schweitzer and Bryan R. Luce made a detailed assessment of the indirect and direct costs attributable to smoking in 1978. Subsequently, Dr. Marvin Kristein of the American Health Foundation updated these figures to correspond to prevailing health care costs in 1980. As indicated in Table 15:1, cigarette-related diseases are responsible for more than $11 billion per year in medical expenses and $36 billion in lost productivity.

Table 15:1. Cost of smoking in the United States per year in January 1980 dollars

	Medical care	Lost earnings	Total
Cancer	$ 1,453,000,000	$ 6,310,000,000	$ 7,763,000,000
Cardiovascular disease	$ 5,120,000,000	$18,230,000,000	$23,350,000,000
Respiratory disease	$ 4,450,000,000	$12,025,000,000	$16,475,000,000
Total	$11,023,000,000	$36,565,000,000	$47,588,000,000
Per adult smoker	$204	$677	$881
Per cigarette sold	$ .02	$ .06	$ .08
Per package sold	$ .36	$ 1.19	$ 1.55

Based on data from Luce and Schweitzer in the *New England Journal of Medicine*, March 9, 1978 and assumptions of 54 million adult smokers and 612 billion cigarettes sold per year.

These figures do not take into account the indirect impact on families, employers, friends, community, etc., or the multiplier effects of lost incomes.*

How tobacco is subsidized

"There is no tobacco subsidy, there is no tobacco subsidy . . ." With these words, North Carolina Senator Jesse Helms exhorted his tobacco state cronies to stand firm in their defense of the tobacco price support system, which they staunchly refuse to call a subsidy.

What is it? According to the U.S. Department of Agriculture, the tobacco price support program is a government-administered program "to stabilize tobacco production and marketing and raise tobacco prices, thereby increasing producer income." It operates through government loans which "provide producers with operating funds, and guarantee them a price at least equal to the support rate." In other words, the federal government lends money to tobacco cooperatives to purchase tobacco that cannot be sold for a designated minimum price. The cooperatives then attempt to sell the tobacco later for a profit. Until recently, money was lent as non-recourse loans, meaning that any losses on these loans were borne by the government, not by tobacco growers.

---

*Editor's note: In 1984, just as this book went to press, a major study was released which estimated that during their lifetimes, middle-aged men who are heavy smokers will suffer an average of $59,000 each in extra medical bills and lost incomes! Published in book form as *The Economic Costs of Smoking and the Benefits of Quitting,* the study was financed by Merrell Dow Pharmaceuticals and directed by medical economist Gerry Oster.

According to Steven Wieckert, legislative aid to Congressman Thomas Petri, who in 1981 sponsored an unsuccessful bill to abolish the tobacco price support program, it is a "badly designed, archaic and feudalistic program that harms those whom it is supposed to help." Mr. Wieckert feels that the program, which is the only farm program not subject to regular review and renewal, is so bad that it will collapse under its own weight within the next 10 to 15 years.

Part of the government tobacco program is an allotment system which limits the amount of tobacco which can be grown each year by requiring that tobacco producers obtain a permit. This limitation of production, unique to tobacco, keeps tobacco prices artificially high, thus ensuring a hefty profit to the owners of tobacco allotments.

The individuals who actually do the farming for their absentee landlords (of which there are many) do not fare so well. The allotment holders receive the lion's share of the tobacco profits, while the tenant farmers generally receive only a small percentage. Yet these are the same small tobacco farmers whose welfare suddenly becomes a great concern to the tobacco industry whenever tobacco is challenged.

The U.S. government also provides tobacco inspection and grading services, a market news service and research and extension services. Total cost to the taxpayer for these government services in 1980 was $28.7 million.

Since the tobacco price support program was initiated in 1933, the government has lost a cumulative total of $57 million in loan principal and approximately $152 million in unpaid interest. Another $200 million was spent on tobacco export subsidies between the years of 1966 and 1972, when the export subsidy was eliminated.

A bill was passed by Congress in 1982 which supposedly made the tobacco price support system self-sustaining. Under this legislation, sponsored by tobacco state legislators, tobacco growers contribute a few cents per pound of tobacco toward maintaining the tobacco price support program. Administrative costs—about $15 million per year—are still borne by the federal government, however.

This legislation, which was pushed through the House and Senate "like lightning" was a reaction to Congressman Petri's bill, which would have eliminated the tobacco price support program and allotment system completely, while increasing the excise tax on cigarettes. Petri's proposed legislation, an indication that the public is becoming increasingly unwilling to lend tobacco a hand, apparently threw enough of a scare into the tobacco advocates to stimulate their own smokescreen legislation.

The contribution

The tobacco industry does make a fairly substantial contribution to the economy if you don't subtract what tobacco costs the nation. It provides employment for more than 400,000 individuals; and total domestic sales are over $20 billion per year including purchases from the fertilizer, advertising, paper and other such auxiliary industries. Exports bring in another $2.2 billion per year. Federal, state and municipal revenues from excise and sales taxes on tobacco products amounted to over $7 billion in 1981.

Net cost

What does it all add up to? So far we have tobacco-related costs to the economy valued at well over $47 billion per year, and direct contributions of tobacco valued at $22 billion per year. Does a $15 billion deficit sum up the full economic impact of tobacco on the U.S. economy?

Not really. Any comparison between the dollar value attributed to tobacco costs in the U.S. and a dollar valuation of tobacco's contribution to the economy would be the proverbial comparison of apples with oranges. No economist has ever attempted a comprehensive and complete assessment of tobacco's effect on the U.S. economy, including both the costs related to tobacco-induced disease and its contribution to the Gross National Product (GNP). Consequently, the "plus" and "minus" estimates which are available have not been derived in the same manner, using the same economic models or taking the same factors into account.

The most comprehensive attempt to assess tobacco's contribution to the GNP was made by the Wharton Applied Research Center at the University of Pennsylvania. This study came up with the rather high figure of $57 billion for tobacco's annual direct and *very indirect* contributions to the economy. Unfortunately, no such comprehensive estimate of the drain which tobacco produces on the economy has ever been made. If it were, the dollar valuation placed on tobacco's economic costs would certainly rise substantially.

Even if one accepts a wide margin of error for both cost and contribution estimates, it is clear that the tobacco industry certainly does not make a vast contribution to our economy as the tobacco advocates would have us believe. Is tobacco's devastation worth such a dubious reward?

Regardless of the exact figure concerning the economic effect of tobacco, the real problem is how the costs and benefits associated with the deadly leaf are dispersed in our society. *What is objectionable is not so much the amounts involved but who benefits and who loses.* The beneficiaries are those who choose to produce, promote or manufacture a patently dangerous product: a few massive tobacco conglomerates, tobacco allotment holders and the auxiliary industries which supply goods and services to the tobacco industry. The burden of payment, however, falls unfairly on everyone who pays taxes, carries insurance coverage or purchases consumer goods.

How nonsmokers pay

The general public is forced to bear the direct costs of smoking in a variety of ways. When an indigent smoker develops cigarette-induced heart disease, his medical bills are likely to be paid by Medicaid or public hospitals, both supported by the tax dollars of nonsmokers. Public taxes pay disability benefits to a smoker disabled by emphysema. When a smoker dies of lung cancer, the general public supports his family through Social Security survivor's benefits.

Thus, although the nonsmoker pays the same amount of taxes as does the smoker, it is the smoker with his much higher probability of contracting and dying from a serious disease who is more likely to benefit from the social and medical services supported by those taxes. When both the smoker and nonsmoker are taxed at the same rate, the disabled smoker receives benefits which are disproportionately higher than his contribution, especially when one considers the fact that the smoker's early death or disability reduces the lifetime total of taxes which he pays into the system.

Tobacco advocates are quick to point out that smokers pay into the system by way of municipal, state and federal excise taxes on cigarettes. However, the amount of revenue obtained from these taxes is woefully inadequate to cover the costs of smoking.

If cigarette smoking provides any economic "benefit" to the general population, it does so by reducing Social Security payouts to smokers who die prematurely. But since cigarette manufacturers cannot admit that smoking kills anyone, they have not yet tried to profit from this argument.

... and pay

Nearly all health insurers and a majority of life insurance companies charge equal premiums to smokers and nonsmokers under both private and group insurance policies. Again, smokers get more than their money's worth when they or their families collect insurance benefits for disease or death. Nonsmokers get less than their money's worth since they pay the same premiums, but aren't as likely to collect the benefits for early death, medical costs or disability.

Smokers also charge their medical bills to nonsmokers through increased consumer prices. It has been estimated that in 1980, smokers spent nearly 150 million more days sick in bed and 81 million more days off the job than did their nonsmoking co-workers. Smokers have also been shown to have a 50 percent greater chance of being hospitalized than nonsmokers; and a recent study found that the job-related accident rate of smokers was twice that of nonsmokers. All of this led to estimated losses in productivity of $36 billion in 1980!

In addition to losses in productivity, employers must bear the expenses of more frequent cleaning and repair of office furnishings and increased costs of air conditioning to filter smoke from the air. Employers who pay all or part of employee premiums in a group insurance plan spend an average of $300 extra per smoker annually. The employer, of course, then passes these costs of smoking to consumers in the form of increased prices.

Fire!

Cigarettes are also a major fire hazard. According to U.S. Fire Administration figures, careless smoking was responsible for over $300 million in fire damage in 1981. Most of these fires occurred in residential buildings, where nearly 2,000 people died and over 3,000 people were injured. The costs of these fires are not usually borne by the smokers who cause them (many of whom burn to death or asphyxiate due to smoke inhalation). Insurance companies pay for the damages—and then pass the cost along to *all* of their property insurance policyholders.

Readjusting the burden

How can the economic burden of tobacco be shifted to where it belongs—on the backs of smokers and tobacco producers? One frequently proposed solution is a "health tax" on cigarettes, with proceeds used to fund social welfare programs which pay for public medical costs of smoking. Such a tax could be graduated according to the relative "safety" of a cigarette, with higher taxes being levied on high tar and nicotine cigarettes and lower taxes on low tar and nicotine-free cigarettes. In 1983, such a tax was proposed by an advisory committee for Social Security as a way to prop up the financially troubled system.

There appears to be an increasing awareness on the part of insurance companies that nonsmokers are much better risks than smokers. As a result, many life insurance companies are willing to give discounts to nonsmokers. The discounts range from a few cents to several dollars per $1,000 of insurance. Even auto insurers are beginning to give discounts to non-smokers, realizing that smokers have more accidents due to the distraction of lighting up while driving and the effects of occupying close quarters filled with cigarette smoke. Health and hospitalization insurance companies have not yet shown much interest in the areas of nonsmoker discounts, but hopefully they will begin to follow the lead of the life insurers in the not-too-distant future.

Litigation against the tobacco industry by victims of cigarette induced illness, discussed in Chapter 16, may also be instrumental in getting the tobacco industry and smokers to pay their own way. If a successful lawsuit were brought against the tobacco industry, paving the way for literally thousands of similar suits, an examination and readjustment of the way in which the economic costs of smoking are allocated might well follow.

Another economic trend favorable to the nonsmoker is the increasing reluctance of many businesses to hire smokers or to allow workers to smoke on the job. Many businesses are also doing their part to help workers stay healthy and productive by conducting or paying for smoking cessation clinics. As more and more companies recognize that smoking adds an unnecessary cost to their operations and begin to take action to cut these costs, they can in turn, cut the cost which the nonsmoking consumer pays for goods and services.

It appears that American business, insurance companies and political officials are beginning to sit up and take notice of the detrimental economic consequences of smoking, and small steps are already being taken toward redistribution of the economic burden. It remains the charge of every nonsmoker, however, to continue to drive toward an eventual equitable distribution of the cost of smoking.

Dr. George Gitlitz, a surgeon who regrets earning money by treating cigarette-related vascular diseases, has proposed a more vigorous plan for encouraging a smoke-free society: 1) federal legislation to mandate insurance discounts to nonsmokers (or raise costs to smokers); 2) legislation to penalize smokers by reducing their salaries and other benefits; 3) a ban on all cigarette advertising, with federal subsidies offered to publications that have become dependent on tobacco ad revenue; and 4) economic incentives to the tobacco industry in the form of educational programs to help tobacco farmers convert to other crops, subsidies for not growing tobacco, and loans and tax breaks to tobacco companies in order to encourage further diversification.

Recommendations for action

As an individual nonsmoker or as a businessperson concerned about company profits, you can take the following steps toward shifting the economic burden of smoking:

- If you are a nonsmoker currently holding a life, health, hospitalization or auto insurance policy which does not offer discounts to nonsmokers, switch to a company that does. Be sure to inform your former insurance company in writing why you made the switch. If enough nonsmokers express dissatisfaction over the way insurance costs are distributed, more insurance companies will give nonsmokers' discounts.
- Write to your congressman, expressing your concern over the economic burden which tobacco places on the U.S. economy and demand that something be done about it.
- If your place of business has no policy or an inadequate policy regard-

ing smoking, *do something* about it. Let the policymakers in your company know how much smoking is costing the company.

• Certainly all industries that require workers to be exposed to hazardous substances which may exacerbate or add to health problems associated with cigarette smoking should immediately ban all on-the-job smoking and direct intensive efforts toward getting current smokers to quit. Such industries should be the frontrunners in adopting a policy of refusing to hire smokers.

16
"Sue the Bastards!"

"The cigarette industry has never paid out one cent as compensation for tobacco-induced injuries. . . . Tobacco law is a defendant's dream come true."—Donald W. Garner,
  Associate Dean
  Southern Illinois School of Law

In this litigious era, when drug, food, automobile and other industries have been ordered to pay out millions of dollars for injuries caused by their products, only the tobacco industry has enjoyed immunity from such obligations. Do you think that an industry that kills far more people per year than all other industries combined should be excused from all responsibility for the health devastation it has been causing?

Suits brought against other industries have usually been for damages to consumers who were using products for constructive or beneficial purposes. Damages have been awarded even for injuries which were unforeseeable or were relatively rare side effects of a product's use. Cigarettes, on the other hand, serve little or no beneficial purpose; and the injuries which they cause are quite foreseeable. With over 100,000 Americans now dying each year from cigarette-induced lung cancer, this problem can hardly be considered a rare side effect.

Since scientific evidence suggesting a link between smoking and cancer was available as far back as the 1930s, it is hard to believe that the tobacco companies were unaware of the dangers of cigarette smoking until they read their own warning labels in 1966. Yet these companies have used a defense of "ignorance" to escape liability in court. Damages have been awarded to consumers who purchased tobacco which contained foreign objects such as worms, fish hooks, human toes, firecrackers, steel particles and snakes, but no lawsuit against a tobacco company based on the inherent health dangers of cigarette smoking has ever succeeded!

What is wrong? Our judicial system has become increasingly disposed to apply strict liability standards to other products or industries which cause harm to consumers or employees. Why has it failed to apply the same standards to the tobacco industry?

Lartigue vs. Liggett & Myers

Only two damage suits against the tobacco industry have ever reached a jury. One was the Lartigue case, filed in Louisiana in 1958. Frank Lartigue, reportedly a "cigarette fiend" since the age of 9, had died of lung cancer in 1955. His widow filed a $779,500 suit for wrongful death against the companies whose tobacco products Mr. Lartigue had smoked: R.J. Reynolds and Liggett and Myers.

An unfortunate blunder by one of Mrs. Lartigue's lawyers caused a mistrial to be called only two days after the case first came to trial. It seems that attorney H. Alva Brumfield had hired a private investigator to find out whether prospective jurors smoked. When several jurors reported being telephoned by the detective pretending to be conducting an academic survey, the judge placed the case at the bottom of the court docket.

The case finally came to trial again in December 1960, with Mrs. Lartigue asking for only $150,000 in damages. Attorneys Brumfield and Melvin Belli charged that the defendants had breached their responsibility to deliver a product which was wholesome, and that they had also failed to warn consumers of the unwholesomeness of their product.

Dr. Alton Ochsner, the prominent thoracic surgeon who had done extensive epidemiological research on the smoking-lung cancer link, testified at the trial. He stated clearly that not only did cigarettes cause Mr. Lartigue's cancer, but that nearly 85 percent of the 2,000 lung cancers he had personally seen were caused by smoking. A deposition by another highly respected lung cancer researcher, Dr. Ernst L. Wynder, was also read before the court. He cited numerous animal studies showing that cigarette-tar condensate had induced cancers when placed on the skin of laboratory animals.

The defense countered with testimony from Dr. Thomas H. Burford, Professor of Surgery at Washington University in St. Louis. Dr. Burford testified that both he and other members of the scientific community remained unconvinced that cigarette smoking was a major cause of lung cancer. After reciting a long list of other ailments from which he believed that Frank Lartigue had suffered, the defense attorney asked Dr. Burford if he could say with certainty that cigarette smoking had caused Mr. Lartigue's lung cancer. Dr. Burford said, "No, I cannot. My opinion is cigarette-smoking does not cause cancer of the lung."

After 17 days of proceedings which produced 20 volumes of testimony, the jury was instructed to find the defendants guilty of breach of warranty or negligence only if they had known, or should have known their product was harmful before Mr. Lartigue contracted his lung cancer. After an hour and 40 minutes, the jury reached its decision: Defendants not liable.

Green vs. American Tobacco

Edwin M. Green began smoking Lucky Strike cigarettes in the early 1920s, at the age of 16. He continued to smoke up to three packs of Luckies a day until 1956 when it was discovered that he had lung cancer. In December 1957, Green filed a $1.5 million lawsuit against Luckies' manufacturer, American Tobacco Company, charging that the company's cigarettes were responsible for his illness. Green's lawyer, Dr. Lawrence Hastings, contended that the American Tobacco Company, in selling its product to the public, had warranted by implication its fitness and merchantability and should be held liable for any damages incurred by breach of that warranty. The case did not end until 1970, 12 years after Mr. Green's death at the age of 49.

When the case first went before a Florida jury in 1961, the jurors concluded that smoking Lucky Strikes did indeed cause Mr. Green's lung cancer, but they awarded no damages to his wife and son. Prior to the date when Green's lung cancer was discovered, the jury reasoned, American Tobacco could not have known "by application of reasonable skill and foresight" that smoking Luckies would cause cancer.

This verdict was appealed, however, on the grounds that Florida law did not require that the defendant be aware of the dangers of its product in a suit involving implied warranty of merchantability. The appeals court ordered a new trial.

During the second trial, the judge instructed the jury that if cigarettes endangered any *important number* of smokers, then there would be a breach of implied warranty of fitness for which the manufacturer would be responsible. Faced with the complex task of deciding whether cigarettes were dangerous to the general public, the jury sided with the defendants.

Other lawsuits

The outcome of the Pritchard vs. Liggett and Myers Tobacco Company lawsuit, which was initiated in 1961, was extremely unfortunate. The federal appellate court ruled that Liggett and Myers could be held liable for selling an unmerchantable product if the smoker suffered physical injury as a result of smoking, but the plaintiff did not choose to pursue this line of prosecution. Rather, the suit was based on the theory that the defendant had been negligent by failing to warn of its cigarettes' danger, and that it had falsely implied that its product was fit for consumption. The jury decided that cigarette smoking did indeed cause the plaintiff's injury, but that he had voluntarily assumed the risk of physical harm. The jury also found that no express warranty had been made that the product was merchantable, so the defendant was not negligent.

An appellate court, however, reversed the lower court's finding that the plaintiff had assumed the risk of harm and ordered a new trial. But the plaintiff dropped the lawsuit at that point, even though the prospects of winning appeared promising.

In Albright vs. R.J. Reynolds Tobacco Co., the defendant had already received payment for his lung cancer from a municipality in a previous accident suit, so the courts did not get to address the merits of the claim.

In Hudson vs. R.J. Reynolds Tobacco Company, another Louisiana suit filed in 1958 by Melvin Belli, the plaintiff failed to allege that the risk of lung cancer was a foreseeable consequence of the company's product. In Cooper vs. R.J. Reynolds Tobacco Co., the plaintiff could not prove that the defendant had misrepresented its product's safety. Finally, in Ross vs. Philip Morris and Co., the company was not held liable because the jury concluded that lung and laryngeal cancer were not a foreseeable consequence of the product's use by smokers.

Numerous other lawsuits have been filed against the tobacco companies, but for one reason or another, they have been dropped or dismissed.

Attorney Melvin Belli recently filed damage suits against two major tobacco companies on behalf of the family of John C. Galbraith, a former smoker who died of lung cancer, chronic obstructive pulmonary disease, and heart failure. Plaintiffs alleged that Mr. Galbraith's many ailments were all caused by smoking. Damages were sought under the strict liability theory: that the two companies should be held liable even if they were not aware of the hazards posed by their product or did not act negligently. Mr. Belli also alleged that the tobacco companies were negligent, failed to warn consumers of their products' dangers and that their actions constituted fraud and deceit. If all had gone well, the case could have gone to trial within a year. The cigarette-land godfathers were not idle, however. They petitioned the court and were granted a change of venue (place of trial) which could lead to as much as a 5-year delay in the proceedings.

These cases illustrate how the tobacco industry has been able to slip through a variety of legal "cracks." Their vast legal and financial resources have allowed them to outmaneuver and at times wear out individual plaintiffs. Does this mean that the tobacco industry will continue to kill smokers without penalty?

Probably not. While in the past, the courts were more likely to apply a negligence standard of liability to damage suits, today most courts are choosing to apply a "strict liability" standard. Strict liability means that it doesn't matter whether a manufacturer was negligent in selling a product which he knew or should have known to be dangerous. Under a strict liability standard, a plaintiff may be awarded damages merely because a defendant caused injuries. Since the scientific evidence is clear, proving in court that a plaintiff's lung cancer was caused by cigarettes should not be too difficult.

In New Jersey, which has fairly liberal liability laws, several lawsuits are pending against tobacco companies. These cases are being managed by a consortium of three large law firms which have sufficient resources and commitment to make it likely that one or more trials will result.

Are warning labels dangerous?

There is, however, an obstacle in the way of damage suits under strict liability standards. Ironically, that obstacle was created by the very forces working to protect the public from cigarettes. "Warning: The Surgeon General Has Determined that Smoking Is Dangerous to Your Health" could prove to be a legal windfall for tobacco companies.

By placing this small warning on each cigarette pack, manufacturers may have exempted themselves from liability for injuries or damages to smokers. Even under principles of strict liability, it is necessary that the product be *surprisingly* dangerous to the consumer. Since warnings have existed on cigarette packs since 1966, many smokers can be considered to have voluntarily assumed that risk, absolving the cigarette manufacturers of responsibility. Of course, lung cancer victims who began smoking prior to 1966 did not have the benefit of a warning label, so the notion of voluntary assumption of risk may not apply to them.

The issue of what constitutes an effective warning may be pertinent. "Dangerous to your health" is a rather vague statement. Skipping meals may be dangerous to one's health, but certainly not to the same extent that overdosing on barbiturates (or smoking cigarettes) would be. It could be argued that since the warnings have never specified that smoking can cause death from lung cancer, heart disease, emphysema or other serious illnesses, they have not been explicit enough to allow smokers to make a fully informed choice. The courts have reasoned that way in similar civil liability suits involving sinus medication and spray deodorant. Even though warnings were provided concerning the danger of kidney damage, in one case, and fire, in the other, the courts ruled that the warnings did not provide *sufficient* information to make consumers fully aware of the products' dangers. The same could certainly be said of the cigarette warnings. It might also be relevant that cigarette companies have steadfastly denied that their products are harmful.

Apart from the issue of vagueness or specificity of warning labels is the issue of responsibility to warn that smokers can get hooked. While tobacco dependency was referred to as "habituation" rather than addiction in the 1964 Surgeon General's report, the 1979 report states that "once the smoking habit is acquired, the stage is set for addictive processes to contribute to the maintenance of the habit." The addictive nature of tobacco was actually noted as far back as the 16th century by the English King James I, who stated in his *Counterblaste* that the smoker "soon becomes so obstinately addicted to it that he would sacrifice every pleasure in life rather than give it up." The fact that 85 percent of teenagers who try a few cigarettes "just to see what they're like" become regular smokers may be evidence of a powerful addictive factor. Even more damning is the observation that three months after stopping, 75 percent of smokers have resumed their habit—the same failure rate observed in heroin addicts.

According to Dr. Donald Garner, associate dean of Southern Illinois University School of Law, "Dependency adds a new dimension to smoking for it greatly increases the likelihood of long-term use, and long-term use is the ticket to early death and disability and disease. The cigarette industry's failure to warn of dependency carries powerful legal implications."

Although no plaintiff has yet filed suit against a tobacco company on the basis of failure to warn of addiction, Dr. Garner feels that the chances of winning such a case are good because of a precedent set by the Texas Supreme Court. A suit was brought by the widow of Glenn Crocker, who had inadvertently become addicted to the analgesic, Talwin. When Crocker was unable to obtain the drug, he substituted injections of the narcotic, Demerol, which ultimately caused his death. The court found for the plaintiff, concluding that the manufacturer had a duty to warn Crocker's doctor of the possibility of addiction.

If a case were to be brought against a cigarette manufacturer on grounds of insufficient and vague health warnings or failure to warn of addiction, however, there is always the possibility that the courts would decide that Congress has preempted the entire field of cigarette labeling under the Cigarette Labeling and Advertising Act of 1965 and the Public Health Cigarette Smoking Act of 1969. The courts could also decide that Congress had not preempted the area, but that the warning label was sufficient since it was mandated by Congress. But in similar cases, the courts have ruled that the existence of a federal labeling law does *not* exempt manufacturers from liability for failing to *adequately* warn of their products' hazards.

Respiratory rape

John Banzhaf points out that cigarette smoking could be considered analogous to statutory rape. Under statutory rape laws, a child or adolescent under the age of consent is considered too immature to fully appreciate the significance and consequences of sexual activity. Therefore, an adult who has sexual relations with an adolescent under the age of consent may be charged with rape, even though the child willingly agreed to such relations.

The same could certainly be applied to cigarette smoking. A high percentage of smokers were below the "age of consent" when they began smoking. If youngsters are too immature to fully comprehend the significance and consequences of sexual activity at that age, can they fully understand the hazards of cigarette smoking? Even if they read the health warnings on cigarette packages or are generally aware that smoking may lead to serious disease in later life, they may be unconcerned about health problems at the "ancient" age of 50. They are also seduced by cigarette advertising that misrepresents smoking as the fun, "with it" thing to do.

It can thus be argued that a decision to smoke, made at an early age is not a voluntary one—and that continuation of the habit into adulthood is not a free choice, either, since by the time adults realize the consequences of their initial decision to smoke, they are already addicted. In my opinion, cigarette manufacturers are actually no less guilty of seduction or rape than are molesters who tempt little girls with candy.

Civil adjudication

Dr. Garner, who reviewed the issue of cigarette manufacturer liability in the *Emory Law Journal* and the *Southern California Law Review*, has also suggested that civil adjudication procedures might be used to force tobacco companies and ultimately smokers to pay their own way. Under this theory, government agencies could sue cigarette manufacturers for the cost of treating cigarette-induced illnesses. Medicare, for example, could sue the appropriate tobacco company for the cost of treating a Medicare recipient's lung cancer. Similarly, Social Security could sue for benefits paid to survivors of a smoker who died of cigarette-related heart disease, a public hospital could recover the cost of treating an indigent cigarette smoker's laryngeal cancer, et cetera.

A precedent for this type of legal action has been set by the federal government. The Federal Aid to Dependent Children program can sue delinquent fathers for welfare costs paid to his family, and the U.S. government may sue private industries for costs incurred in cleaning up their oil spills.

Garner proposes that proof of causation could be waived in such cases, due to the vast amount of damning evidence linking cigarette smoking with disease. Evidence that a certain individual had smoked two packs of cigarettes per day for 20 years would be introduced as presumptive "proof" that his lung cancer was caused by cigarettes. The Black Lung Benefits Act of 1972 requires coal mine operators to compensate workers who develop "black lung." Under this Act, proof that a coal miner's disease was caused by working in the mines is established by presumption. In other words, if a man who worked in a coal mine for 10 years develops "black lung," it is presumed that his work was the cause of his illness. It then falls upon the mine operator to rebut this presumption. There is no reason why the same procedure could not be applied to cigarette manufacturer liability cases.

In cases where one particular brand of cigarette could not be identified as the cause of a smoker's disease, the courts could apply the same principle used in awarding damages in lawsuits involving diethylstilbestrol (DES), assigning liability based on a cigarette manufacturer's market share. To reduce the burden on the court system, Dr. Garner also suggests that a special administrative tribunal composed of technical experts could handle the cases, in a manner similar to the workmen's compensation boards. While courts might theoretically announce civil adjudication liability, it is more likely that new state or federal laws would be required to implement the type of system proposed by Dr. Garner.

Asbestos and tobacco

The tobacco industry does face challengers in the legal arena who are as big, wealthy, resourceful and determined as the tobacco companies: the asbestos industry and its insurance carriers.

Since asbestos workers in Tyler, Texas won their case in 1975, the asbestos industry has been inundated with lawsuits. Thousands of insulation workers, shipyard workers and others have developed asbestosis (a lung disease similar to emphysema), lung cancer or mesothelioma (a rare tumor of the lining of the lung) as a result of years of exposure to asbestos. Since it is nearly impossible to determine which of the over 200 asbestos companies' products a given worker was exposed to, suits have been brought against multiple defendants. Most suits have been settled out of court, but many which have gone to a jury have been won by the plaintiffs. Since thousands of asbestos-related illnesses occur *each year*, the amount which the asbestos industry may be required to pay for damages is astronomical.

As the asbestos industry has begun to feel the pinch of the tremendous costs of litigating claims, it has begun to challenge others to share these costs. Asbestos companies are charging that the U.S. government should be held accountable for exposing over 4 million workers to asbestos in Naval shipyards during World War II. And the tobacco industry has also been recognized as a leading player in the asbestos tragedy.

Lung cancer accounts for about 20 percent of deaths in workers heavily exposed to asbestos. According to a major investigation involving over 17,000 asbestos workers: 1) nonsmoking asbestos workers have five times the risk of dying from lung cancer as nonsmokers who have not been exposed to asbestos; 2) smokers not exposed to asbestos have 10 times the risk of dying from lung cancer as nonsmokers not exposed to asbestos; 3) asbestos workers who do smoke are more than 50 times as likely to die of lung cancer as nonsmoking members of the general population; and 4) asbestos workers who smoke a pack or more of cigarettes per day have 87 times the general nonsmokers' risk of lung cancer death!

Lung damage caused by cigarette smoking can also increase disability and likelihood of death from asbestosis. Thus, cigarette smoking is at least partially responsible for a significant proportion of the asbestos-related disease cases which have resulted in lawsuits. Dr. Irving Selikoff, a noted expert on asbestos disease, estimates that as many as 50 percent of future asbestos-related claims might be avoided if workers already exposed to asbestos would stop smoking!

In an attempt to reduce their losses and highlight tobacco's role in the asbestos tragedy, asbestos companies have used the "empty chair" defense. Pointing to an empty chair, they claim that tobacco companies should occupy that chair as a defendant, since they are totally or partially responsible for the plaintiff's injuries. While plausible, this defense has not been particularly successful.

There is, however, another way to try to force the tobacco industry to pay its share. That method is the filing of a cross-complaint, a legal device for bringing another defendant into a lawsuit. The one asbestos manufacturer which has filed such a cross-complaint, Standard Asbestos Manufacturing and Insulating Company, did so at the behest of its insurer, Boston-based Commercial Union Insurance Companies.

Filed in 1980, the cross-complaint alleged that cigarettes caused and contributed to injuries and damages suffered by several hundred asbestos workers who were suing. Standard Asbestos was later dropped from the lawsuit for reasons unrelated to the cross-complaint, so this strategy did not get tested in court.

Commercial Union, which holds primary and excess coverage for many companies involved with asbestos, plans to file more cross-complaints once suitable cases are found. However, other asbestos and insurance companies have been reluctant to follow Commercial Union's bold lead. John Banzhaf suggests that other companies are afraid to "tangle with the big boys." Given the tobacco industry's enormous and widespread clout, such a fear may be justified. In fact, when Commercial Union approached the Tobacco Institute to discuss the tobacco industry's involvement in the issue, the Institute's only response was a thinly veiled threat, intimating that if Commercial Union was stupid enough to sue the tobacco industry then it was free to do so.

David Pullen, manager of U.S. Government Affairs for Manville Corporation, gave another reason for the asbestos companies' lack of enthusiasm to take on the tobacco industry. So far, asbestos companies have won about half of the lawsuits. But a plaintiff is more likely to win if both parties accused of being responsible for his injuries accuse each other. With tobacco in the courtroom, the odds of successful defense are reduced, says Mr. Pullen. "The jury may split the award down the middle, saying that we know this guy is sick, but we can't tell how much is due to asbestos and how much is cigarettes."

Other observers feel the main reason that asbestos and insurance companies aren't going after tobacco is related to legal costs. The tobacco industry would spend almost any amount of money to avoid losing one case, since this would open the way for many more lawsuits. Most asbestos and insurance companies don't want to risk having to match those expenditures.

Mr. Banzhaf also believes that at least some of the insurance carriers for the asbestos companies may also insure tobacco companies. In such cases, the insurance company has nothing to gain by bringing a tobacco company into court.

Commercial Union's plan to file more cross-complaints against the tobacco industry still appears promising. What are the chances of winning such a suit?

John M. Pinney, former Director of the Office of Smoking and Health, says "The situation is difficult to put a finger on. There's a fair chance of having a jury who already believes that cigarette smoking caused lung cancer, but there's also a fair chance of getting a jury which believes that someone exposed to large amounts of asbestos deserves full compensation from the asbestos industry."

Mr. Banzhaf, who is currently looking into ways to *force* the insurance and asbestos companies to file cross-complaints against the tobacco industry, thinks the chances of eventually winning are fair to good. He believes that there is a significantly greater probability of winning under these circumstances than in past lawsuits filed against the tobacco industry alone. He said recently:

> The major difference is that previous single-plaintiff cases have been grossly underfinanced and not willing or able to go the distance. The tobacco industry is willing to spend $500,000 to win a $100,000 case, but so is the asbestos industry. This situation is not really different in a factual or legal sense. Rather the tobacco industry will be faced with an opponent which is as big, bad and well-financed as itself.

Where there's smoke . . .

The best opportunity to open a crack in tobacco's legal dam may not be a suit filed by a lung cancer victim, but one filed against a cigarette manufacturer on behalf of someone injured or killed by a cigarette-ignited fire, says Banzhaf. Each year 2,000 people die and 3,000 more are injured in cigarette-related fires. The victims include smokers, their families and innocent occupants of the smoker's hotel or apartment building.

Preferably, says Banzhaf, the suit should be filed on behalf of an innocent bystander such as a neighbor who was injured or killed by the cigarette-induced blaze. This would avoid the problem of contributory negligence by the smoker who started the fire. Such suits could be filed against cigarette manufacturers on the basis of negligence in marketing a dangerous product (i.e., a known fire hazard) or under standards of strict liability. While the cigarette manufacturer might counter that cigarettes, like certain drugs or vaccines, are an "unavoidably dangerous product," this would be a highly questionable assertion.

It is not difficult to manufacture self-extinguishing cigarettes which would be far less likely to start a fire when left unattended. In fact, the Consumer Product Safety Commission (CPSC) once attempted to force cigarette manufacturers to produce a self-extinguishing cigarette when it first uncovered the astonishing accidental death and injury figures due to cigarette-ignited fires. However, Congress, then considering the 1972 Consumer Product Safety Act, quickly removed tobacco products from CPSC jurisdiction.

The cigarette industry has resisted the development and marketing of self-extinguishing cigarettes since such cigarettes would have to be re-lit every time the smoker laid one down for a few minutes—an inconvenience which might deter them from smoking so automatically. More important, in all likelihood, is the fact that if cigarettes didn't burn up so quickly, fewer cigarettes would be consumed, leading to lower profits for the industry.

Cigarette manufacturers actually *add* substances to prevent cigarettes from self-extinguishing. And they will no doubt continue to do so until a plaintiff wins substantial damages for injury or death resulting from a fire caused by their product. Banzhaf feels that such a suit would be easier to win than a cigarette-lung cancer suit, due to the more obvious relationship between cigarettes and fires. (Although, given the tobacco industry's history of outrageous defenses of their product, it would not be surprising if the Tobacco Institute began asserting that the relationship between cigarettes and fires was purely statistical!)

The tobacco Goliath

If a single lawsuit against the tobacco industry by a lung cancer victim or even the asbestos insurance carriers were to succeed, the implications for the tobacco industry would be grave. One successful suit would open the way for thousands of similar suits, as the asbestos industry's experience has shown. However, it can be anticipated that if the cigarette industry were to be found liable for damages caused by its products, the industry would act quickly to protect itself. "If the bill for medical expenses, wages, and pain suffered by one out of a thousand smokers were laid at the feet of the cigarette industry, it's likely that the industry would seek and find Congressional immunity," writes Dr. Garner.

Congressional intervention might not result in the tobacco industry getting off scot-free, however. There would undoubtedly be intense public scrutiny of the way in which the economic costs of smoking are divided between non-smokers and smokers. In return for Congressional immunity from future civil liability suits, the tobacco industry would probably be forced to accept some kind of a law requiring it (and ultimately, the smokers themselves) to pay for the enormous costs of smoking rather than dispersing these costs throughout the general population.

Such a readjustment of the economic burden of smoking could be most efficiently accomplished by imposition of a "safety tax" on cigarettes. This tax could be designed so that cigarettes with high tar and carbon monoxide levels would be more highly taxed than those with low tar and carbon monoxide levels. (This tax could, of course, also apply to any other dangerous components of tobacco combustion.) An added bonus of such a tax would be the incentive which it would provide to cigarette manufacturers to produce safer cigarettes. Consumption of the safer cigarettes would then further reduce the extent of cigarette-induced illness for which the public must pay. In the absence of a graduated "safety tax," uniformly higher taxes could be imposed on all cigarettes, with proceeds earmarked to fund increased social welfare spending.

Perhaps an even more important role which any kind of successful litigation against the tobacco industry might play would be the damage which a legal loss would do to the industry's image of invulnerability. Currently, everyone seems to be running scared from what is perceived to be an invincible Goliath. A successful lawsuit might awaken health groups, political officials, big business, the media and smoking and non-smoking individuals to the fact that the tobacco industry, just like any other, can be held accountable for all the trouble it causes.

17

Exploitation of Developing Countries Should be Ended

"We recognized early that ours is a global business and built markets around the world. Our future is particularly bright in developing areas, where income and population are growing."—Joseph Cullman III, Chairman, Executive Committee Philip Morris, Inc., 1983

Faced with the prospect of dwindling sales in the United States, Great Britain and other developed countries, the tobacco industry sought new means of preserving its economic health. Diversification into such areas as Chinese foods, beer and shipping had already begun as the tobacco companies had anticipated how increased public awareness would lessen cigarette sales. Loyal to their original roots, however, the companies continued to seek new markets for cigarettes.

The adult male market was pretty well saturated, women had already "come a long way," and youth market was—at least ostensibly—off limits. Where could the smoking gun find its next target?

The perfect victim

Fortunately for the tobacco industry, the perfect victim lay waiting: the Third World. More than half of the world's population of potential smokers was contained in the Third World countries of Africa, Latin America and Asia. These peoples were extremely poor but were also quite eager to mimic the customs enjoyed by their richer neighbors. Farmers would welcome the opportunity to increase their incomes by growing tobacco, and governments would find it difficult to refuse the lure of tax revenues from the sale and export of cigarettes. Finally, financially strapped governments would be unlikely to spend their limited funds on programs to regulate tobacco marketing and advertising or to educate their people on the hazards of cigarette smoking.

The tragedy

The tobacco industry has been quite successful in turning the Third World into "Marlboro Country." Between 1970 and 1980, per capita cigarette consumption rose less than 4 percent in North America. In Africa, however, it increased 32 percent, while in Latin America it increased 24 percent.

Thus it appears that the Third World has been consuming an increasingly higher percentage of the world's tobacco products. These ever-increasing amounts have been coming from two sources: imports from the major industrialized tobacco-growing countries, and locally grown tobacco produced with encouragement from the multinational tobacco companies.

Judging from the variety of companies listed on the cigarette packages, the players in the Third World tobacco caper might appear to be the proverbial cast of thousands. However, seven multinational companies actually dominate the situation. To give the illusion that a tobacco monopoly doesn't exist in any given country, they often change and exchange brands and company names.

British-American Tobacco (BAT) ranks number one in world domination of the tobacco market. Other multinational companies which figure heavily in the Third World are Philip Morris, R.J. Reynolds, American Brands, and Universal Leaf Tobacco in the U.S., and Imperial and Rothmans in Britain. BAT was the first company to get into the Third World market, but it soon lost its monopoly in several areas when the U.S. companies joined suit in 1976. Together, these seven companies control 75 percent of tobacco production in the "free world."

There is another villain whose name you won't find on the cigarette packages consumed in the Third World: the United States government. Through price supports, export assistance and donations of tobacco for needy countries, the U.S. government has played an important role in encouraging the Third World to take up the deadly habit.

"Tobacco for Peace"

Exports and donations of agricultural products to developing nations have traditionally had a dual purpose. They provide a handy way for agricultural producers to get rid of their surpluses, while making much-needed food available to the hungry nations of the world. However, with respect to tobacco, the benefits have been reaped only by the givers.

In 1954, tobacco became eligible for inclusion in America's Food for Peace program. Under this tax-supported program, the United States Department of Agriculture shipped between $17 million and $66 million in tobacco products per year, along with food, to the hungry countries of the world. (How tobacco can help feed a starving population has never been explained.) In response to serious criticism from international health officials, this practice was abandoned in 1980. By this time, however, the exports had achieved the aims of the tobacco industry. Hungry people all over the world had developed a new appetite—for cigarettes.

The World Bank

The Washington-based World Bank touts a commitment to help the "poorest of the poor" raise their living standards. It has pledged to increase its investments in health and energy-related projects in the Third World. Yet it has also played a major role in promoting tobacco. Using funds garnered from taxpayers in the United States and other developed countries, the bank has lent millions of dollars to countries such as Zambia, Malawi, and Tanzania to support increased tobacco production.

The bait

Why would a government that cannot grow or import sufficient quantities of food for its population, or a farmer who can barely raise enough to feed his family, want to spend scarce money on imported cigarettes or divert precious farmland to the growing of tobacco?

Governments are lured by the promise of substantial revenues from taxes on tobacco products. In Brazil, for example, they generate nearly 12 percent of the government's revenue. But these taxes also provide a convenient method whereby the tobacco companies can make governments dependent on them. A dependency on tobacco taxes plus the fact that friends and relatives of government officials are often made "directors" of the local tobacco subsidiary, ensures that the government will look out for the interests of the tobacco company.

Governments are also enticed by the prospects of becoming tobacco exporters, thereby improving their balance of trade. Third World countries desperately need foreign currency to buy commodities from abroad, and tobacco is presented as an easy solution to this problem. Actually, as discussed below, the value of tobacco in providing local and foreign revenues is highly exaggerated.

Farmers are offered all the assistance they need to convert to the growing of tobacco. A field staff comes in to show the farmer how to plant, tend and harvest the plants. The tobacco companies may also administer and guarantee loans to farmers from the local government. Since farmers usually must sell directly to the tobacco companies, they are also tempted by the promises of prompt payment and comparatively high prices. But farmers, too, are in for a surprise.

That special taste of success

"The smoking habit does not increase spontaneously; it has to be created." Such is the opinion of Gamini Senevitatne, writing for WHO, who understands why the tobacco industry spends over $12½ billion per year to encourage people to smoke. Yet the industry denies that its massive worldwide advertising efforts are aimed at persuading people to *start* smoking. It claims that advertising doesn't endorse smoking per se, that it merely influences smokers to buy a particular brand. But Third World ads indicate otherwise:

"555 State Express: That Special Taste of Success."
"He's a College Man. He smokes Varsity."
"Trust a Graduate."
"The cigarette for the V.I.P."

These slogans, accompanied by pictures of well-dressed, sophisticated smokers are obviously designed to persuade the poor people of the world that smoking is a mark of success, wealth, and social prestige.

Advertising messages are transmitted throughout the developing world in many, often ingenious, ways. In Kenya, for example, where BAT has a monopoly, its mobile cinema provides free movies (complete with cigarette commercials) to over one million prospective customers per month. Advertising posters are not allowed in Kenya, but cigarette distributors paint their houses with the colors of the cigarette packages which they market. In Ghana and Malaysia, tobacco companies sponsor many sports and social events.

Cigarette marketing and distribution efforts in the Third World are certainly heroic, if nothing else. The tobacco companies, through a vast network of land rovers, bicycles, donkeys, and even camels, ensure that even the retailer in the most remote area of his country will receive his weekly allotment of cigarettes, which may only be one pack.

Cigarettes are sold by old ladies in market stalls and by school children on the streets. They are often purchased one or two "sticks" at a time because the price of an entire pack may be out of the reach of the average smoker. Cigarette sales are especially high on paydays.

American-made brands are heavily promoted in the Third World. Not surprisingly, they are also in great demand. *World Tobacco*, the trade journal of the tobacco industry, explains this phenomenon as follows:

> There is an increasing inclination around the world, in both developed and underdeveloped countries, for personal sophistication to imply internationalism of outlook. Perhaps the cheapest way for someone climbing the social ladder to assert his international mindedness is to smoke an international cigarette.

While the success which supposedly accompanies cigarette smoking has eluded Third World smokers, it has come quite easily to the tobacco companies. In Malaysia, where men have been encouraged to believe that smoking is a sign of high wages and being a "man about town," cigarette consumption increased 120 percent between 1967 and 1979. Sixty percent of Malaysian males over 15 years of age are now "men about town."

One reason for the apparent success of cigarette advertising in the Third World may be that its consumers have not been subjected to enough advertising messages to have become generally skeptical toward advertising. Another factor may be that they have little else to believe in.

Full-flavored cigarettes

Adding to the peril faced by Third World smokers is the fact that the cigarettes which they consume often contain twice as much cancer-causing tar as do cigarettes of the same brand sold in the industrialized world. According to War on Want estimates, while a Kent cigarette sold in the United States in 1977 would deliver only 15.5 mg of tar, one sold in the Philippines would deliver 33 mg of tar.

According to an article in *World Tobacco*, having low-tar versions beside the "bolder" versions would cause confusion detrimental to the sales of both because Third World citizens don't understand the "low-tar argument." The article does not state whether anyone has ever attempted to explain the "low-tar argument" to Third World smokers. A BAT official offered another enlightening explanation to War on Want's Mike Muller. Since Third World smokers can afford only a few cigarettes per day, they want a strong cigarette "which they can really enjoy."

While 95 percent of the developed nations of the world have laws pertaining to cigarette marketing and health warnings, only 24 percent of the underdeveloped countries have such regulations. In 1978, WHO recommended that all countries mandate health warnings on cigarette packages, stop cigarette promotion, and develop national policies toward prevention of smoking. These recommendations have largely been ignored by the Third World countries. While a few countries, such as Malaysia, require vague health warnings on cigarette packages, others, such as Kenya, require no health warnings whatsoever. In Taiwan, instead of a health warning, the side of each cigarette package bears a morale booster. Consequently, few smokers in the Third World are aware of the dangers to which they are exposing themselves every time they light up a cigarette.

The obvious consequences

When the tobacco companies began their aggressive push in the United States and Europe in the 1920s, disastrous health consequences of the habit were not known. We may therefore take a somewhat lenient view of their early advertising and marketing ploys, for they were truly ignorant of the dangers of their product. But the companies should not be so pardoned for their intensive marketing and advertising compaigns in the Third World countries. Their well-planned assault began despite incontrovertible evidence that smoking kills people.

Spokesmen for the tobacco industry will, of course, disagree with that notion. In an interview with the War on Want's Mike Muller, BAT's Richard Haddon rationalized that because Third World smokers smoke fewer cigarettes than do smokers in industrialized countries, there is less cause for concern about health problems. "Even medical men say that you have to smoke a certain number of cigarettes—perhaps in excess of 15 a day—to possibly be at risk from smoking."

This statement is contrary to scientific evidence which shows that smoking any number of cigarettes increases the risk of premature death. It is also refuted by epidemiological evidence that smoking-related diseases such as bronchial and laryngeal cancer, emphysema and heart disease have been increasing in Latin America, along with cigarette smoking. In Brazil, where about 135 billion cigarettes were smoked in 1981, smoking-related diseases far outnumber infections as the leading cause of death. In India, where per capita cigarette consumption is still very low, epidemiological studies on large groups of people have shown an association between smoking and heart disease and chronic bronchitis. Esophageal cancer rates in Durban, South Africa, and in Rhodesia are now among the world's highest. Lung cancer was virtually nonexistent in East Africa until recently. So was cigarette smoking.

Governments now value the economic benefits associated with tobacco sales. But it remains to be seen whether tobacco will stay in favor in 10 to 20 years when these governments have to bear the brunt of health care costs for those afflicted with lung cancer, coronary heart disease and emphysema. While health care for victims of these diseases might not be so elaborate or available as that in the United States, these incapacitated people will have to be cared for in some fashion. It is virtually certain that the cost of that health care, combined with the costs of lost workdays, decreased productivity and fire damage and injury, will substantially diminish, if not surpass the apparent profits gained by tobacco taxation and exporting in the developing countries, just as it does today in the United States.

Of course, some government officials in these countries look at the bright side of things. "So you stop people dying—what do they do then?" remarked Raymos Lyatuu, Tobacco Authority of Tanzania's General Manager, to Mr. Muller. In other words, it may be cheaper to kill "excess" people with cigarettes rather than worry about supporting them.

So it appears that the Third World will provide an experimental population for a gigantic prospective epidemiological study. And as lung cancer, heart disease, emphysema and other smoking-related diseases escalate dramatically, tobacco companies will give their predictable rationalizations for the evidence: "Improved diagnosis, genetic susceptibility . . . all statistical."

The not-so-obvious consequences

Aside from the obvious direct health consequences of smoking which the developing countries must face, tobacco takes its toll in more subtle, but no less insidious ways. Although food is the commodity which these countries need most, millions of acres of farmland are being diverted to the production of tobacco, necessitating more food imports.

The tobacco companies insist that tobacco farming is not detrimental to food crop production, and, in fact, even benefits it by being a "teaching crop." But they don't really explain why farmers will learn more by growing tobacco than by growing cotton or groundnuts. Tobacco requires short bursts of intensive activity in its production—which often deprives food producers of the labor they need at critical times. Tobacco farming also leaves a large number of people unemployed during most of the year.

Tobacco will take its toll on the environment in many developing countries. One of the major steps in the processing of tobacco is a procedure known as flue-curing, in which the tobacco must be kept at high temperatures for about a week in order to cause the fermentation reaction that produces the characteristic yellow of the leaf. This is an energy-intensive process. While the industrialized countries can afford to use oil and gas for this purpose, most underdeveloped nations cannot. Their major source of energy is still wood. Wood is used up at the rate of one acre of forest per acre of flue-cured Virginia tobacco or one tree per 300 cigarettes.

The supply of woodlands is not limitless. According to a 1977 report of the United Nations environmental program, the firewood shortage may soon become the "poor man's energy crisis." Eastern Kenya, Pakistan, and even heavily-forested Brazil have already begun to suffer the consequences of wanton use of firewood to cure tobacco. In those countries, tobacco farmers are now experimenting with the use of coal, solar energy and expensive imported fuels to keep the flue-fires burning.

Also, since tobacco flourishes in sandy soils, it is often grown in areas bordering on deserts in the Third World. As trees in these areas are cut down to supply wood for flue-curing, the process of desertification is accelerated and farmers are forced to relocate in less arid regions, where tobacco replaces food crops.

So, while the transnational tobacco companies are accruing impressive profits from planting tobacco in the Third World, the countries themselves are harvesting unemployment, hunger, and desert land.

Pipe dreams

Every Third World country entering into tobacco production dreams of being able to export tobacco. But there is actually little chance of improving their balance of trade in this manner because Third World countries themselves are the intended consumers of the tobacco they produce. Tobacco companies did not venture into the Third World because they lacked the land or resources for tobacco production at home, but because they lacked a sufficient market.

If current trends continue, the Third World will consume a progressively greater share of the world's tobacco products and the rich countries, to which they hope to export, will consume less and less. Apparently, someone forgot to tell the Third World nations that they were getting into the tobacco business a few decades too late!

Even if they have the tobacco to export, it is very difficult for developing nations to produce processed exports. They simply cannot afford to import the paper, packaging materials and machinery needed for the manufacture of cigarettes. Often the tobacco companies control the supply and price of these materials.

They also control the prices which are paid for Third World tobacco crops. Compulsory sales agreements are common in developing countries. The farmer must sell his tobacco to the company, at the company's price, which often happens to be substantially lower than its value in the international market. Thus many farmers who begin growing tobacco with the hope of achieving financial independence end up enslaved by a tobacco company.

While the tax revenues which the tobacco industry provides to the developing countries may be real, the income from such taxes is often significantly reduced by smuggling and bootlegging. Diversion of tobacco profits into the pockets of government officials is not unheard of, either.

All of these hidden costs of tobacco production result in a net profit of close to zero for many developing countries. As bad as the situation is now, it can only become worse. For while the tobacco companies were encouraging the Third World to get into the tobacco game, they were changing the rules.

In the developed world, new technology is making tobacco production much less labor intensive, and more economical. It is also making tobacco and tobacco product exports even more competitive in the world market. Mechanical harvesters reduce field labor by 85 percent, and bulk barns slash curing labor by over half. New methods of "reconstituting" tobacco have increased the use of previously unacceptable tobaccos, thus reducing dependence on foreign producers.

This new technology is not yet suitable for the Third World. Mechanical harvesting only becomes economical at 30 acres, and it is a rare Third World farmer who has a plot anywhere near that size. Chemical reconstitution processes are designed for large, sophisticated factories, not the tiny, primitive plants in the Third World. Third World tobacco producers will thus continue to utilize outmoded, labor-intensive methods of growing and processing, and their tobacco products will become less and less competitive in the export market.

Eventually, the Third World countries will have to switch to new technologies—maybe in a generation or so. The costs of doing this will be high. Jobs will be eliminated. Reconstituted tobaccos will replace those "high-quality" tobaccos which the Third World farmer foolishly believes to be indispensable to particular cigarette brands.

It seems that the people of the Third World have consigned themselves to face disease, hunger, destruction of precious resources and even economic loss, all for the sake of pipe dreams.

Recommendations for action

Is this grim scenario of death, ill health and depletion of natural resources inevitable? Should Americans sit by idly while Third World assets go up in a puff of tobacco smoke? Although the picture of a tobacco-addicted Third World appears bleak, if the appropriate agencies, organizations and governments take action immediately, there is certainly still hope of reducing, if not eliminating, tobacco's toll.

First, and foremost, the United Nation's Food and Agriculture Organization (FAO) should immediately cease its current policy of helping developing countries learn to grow tobacco. It should also stop helping the tobacco industry invade the Third World—something, of course, which FAO will not admit to doing. Any developing country requesting assistance in growing tobacco should be encouraged by FAO to grow an alternative crop which can be used to obtain foreign currency or feed its population. *Under no circumstances* should FAO assist any country in growing tobacco, even if it appears to be economically advantageous to do so. The short-term gains which might accrue from tobacco production are certainly far outweighed by the disastrous long-term effects on the resources of a developing country. Crop diversification should be strongly encouraged, since other lucrative cash crops would probably be the strongest possible disincentive to tobacco production. The FAO should also cooperate with the educational aims of the World Health Organization of the United Nations (WHO) by informing Third World governments of the ecological, agricultural and economic consequences of tobacco production.

Although the United States government does not directly provide export assistance to American tobacco companies at the present time, it does assist them indirectly to export the deadly weed to developing countries. The government's program of direct credit to exporters was discontinued in 1980, but was replaced by a program of credit guarantees to commercial lenders financing tobacco export sales in what appears to be another one of tobacco's "now you see it, now you don't" magic tricks. Our federal government should divorce itself totally from tobacco exports to developing countries, or any other country for that matter, along with the elimination of government involvement in any kind of tobacco price support system. *It's time for the American taxpayer to stop subsidizing death, deforestation and ill health in the Third World.*

One positive way in which the U.S. government could become involved in the Third World smoking issue is to mandate that *all* cigarette packages carry a health warning, not just those intended for domestic consumption.

Currently the governments of the United States and many other developed countries permit the exporting of cigarettes without any health warnings, while requiring warnings on the cigarette packs purchased by their own citizens. This seems a tacit statement that the health and welfare of Third World citizens is less important than that of the residents of their own country. Can any country which even gives lip service to humanitarian ideals and goals continue to allow this type of deadly discrimination?

All cigarette packages targeted for export should be required to bear the same warning labels that domestically consumed cigarettes carry, *printed in the language of the intended export country*. While the efficacy of warning labels in preventing or deterring smoking is questionable, a weak warning is still better than no warning at all!

The World Health Organization recommends that legislative action restricting smoking and cigarette promotion be implemented in all developing countries. According to Ruth Roemer, writing in the WHO publication, *Legislative Action to Combat the World Smoking Epidemic*, legislation would not only speak loudly for the government's official policy towards smoking, but it would also serve as a basis to educate young people on this vital issue, to disseminate information to the general public, to promote smoking cessation activities, and to encourage a smoke-free environment.

WHO, unfortunately, can only *suggest* such measures to the developing countries, and given the strong economic interest which the governments of these countries have in tobacco, adoption of tough anti-smoking legislation in the Third World seems unlikely in the near future. WHO is also hampered by the contradictory FAO activities and by a lack of firm support by the United Nations in general, so its current efforts, while admirable, may be largely futile.

Religious organizations can provide personnel to attack the Third

World tobacco problem. Most denominations are not actively involved at present, but if those committed to humanitarian and missionary work become sufficiently concerned, they could become very effective in the war against tobacco. So could international volunteer organizations, such as the Peace Corps, which could provide public health education and teach Third World farmers how to grow substitute crops.

If you are concerned about the Third World tobacco tragedy you can do the following:

• Encourage your Congressional representatives to support legislation mandating warning labels on cigarette packages exported to the Third World as well as legislation divorcing the U.S. government itself from any export assistance to the tobacco industry.

• If you are affiliated with an international volunteer organization, urge it to get involved in the Third World tobacco issue.

• If you belong to a religious denomination involved in missionary or other international charitable work, bring the problem of tobacco in the Third World to the attention of the policymakers within your denomination. Urge them to get involved in this highly important issue before it's too late.

18

Cigarette Advertising Should be Stopped

We are learning what an industry fighting for survival is capable of doing to erase an unfavorable image. . . . With its self-generating wealth . . . the tobacco industry has curried favor with legislatures, infiltrated papers with lush advertising fees and enticed them into publishing planted stories and kept legal, ethical and medical challenges at bay with an array of hired talent . . . aiming its propaganda at the most impressionable years and tainting a long delayed social reform by identifying itself as a symbol of feminine independence.-Dwight Bollinger, in Language: the Loaded Weapon

The Cigarette Advertising and Labeling Act of 1971 was passed with the hope that eliminating cigarette ads from television and radio would reduce the extent of cigarette advertising. This hypothesis was about as realistic as the concept that closing one of many doors of a lion's cage will prevent the beast from charging out to attack its prey.

It might be argued that the broadcast ban did protect young children who presumably encounter more ads on the electronic media than in magazines and newspapers. But of all companies that regularly advertise their products, few can match the persistence and lavishness of cigarette manufacturers. During the first year after the broadcast ban, money spent on cigarette advertising on billboards, posters, transit ads, and the like jumped over 5-fold. Within ten years, tobacco firms were spending over a billion promotional dollars per year (see Tables 18:1 and 18:2). Although the current ads are tamer than those of George Washington Hill's days when advertising truly "went mad," they are still the most blatant, widespread, dangerous and deceptive in the marketplace.

Most of today's ads emphasize vitality with suggestions of health, outdoor activity, femininity, romance, pleasure and relaxation. Young people are shown bobsledding, taking a smoke after a swim or tennis, whooping it up at an All-American ice cream parlor. A lovely girl in a country setting

Table 18:1. Cigarette Advertising and Promotional Expenses 1970 to 1979 (Thousands of Dollars)

Type of Advertising	Before Cigarette ads were banned from TV/Radio 1970	% of TOTAL	After Cigarette ads were banned from TV/Radio 1975	% of TOTAL	1976	% of TOTAL	1977	% of TOTAL	1978	% of TOTAL	1979	% of TOTAL
Newspaper	14,026	3.9	104,460	21.3	115,808	24.4	190,677	24.5	186,947	21.4	240,978	22.2
Magazine	50,018	13.9	131,199	26.6	148,032	23.2	173,296	22.2	184,236	21.1	257,715	23.8
Outdoor	7,338	2.0	84,329	17.2	102,689	16.1	120,338	15.4	149,010	17.0	162,966	15.0
Transit	5,354	1.5	10,852	2.2	19,341	3.0	21,530	2.8	22,899	2.6	21,151	2.0
Point of Sale	11,663	3.2	35,317	7.2	44,176	6.9	46,220	5.9	57,384	6.6	66,096	6.1
Promotional Allowances	33,789	9.4	72,018	14.7	82,523	12.9	108,227	13.9	125,148	14.3	137,111	12.7
Sampling Distribution	11,775	3.3	24,196	4.9	40,390	6.3	47,683	6.1	47,376	5.4	64,286	5.9
Distribution Bearing Name	2,649	0.7	6,775	1.4	9,847	1.5	24,636	3.2	32,673	3.7	44,839	4.1
Distribution Not Bearing Name	3,012	0.8	3,313	0.7	10,183	1.6	11,161	1.4	15,608	1.8	17,190	1.6
Special Events	544	0.2	8,484	1.7	7,946	1.3	9,538	1.2	11,590	1.3	10,783	1.0
All Others	220,841	61.1	10,311	2.1	18,182	2.8	26,157	3.4	42,100	4.8	60,310	5.6
TOTAL	361,000	100.0	491,254	100.0	599,117	100.0	779,463	100.0	874,972	100.0	1,083,425	100.0

Table 18:2. 1981 Cigarette advertising revenue

Magazine	Pages of cigarette ads per year	Yearly revenue	Percent of cigarette ads
Atlantic	42.0	374,724	20.5
Better Homes	194.5	12,945,229	13.5
Black Enterprise	36.0	370,859	7.8
Bon Appetit	64.8	1,044,150	6.8
Book Digest	16.8	95,520	6.5
Business Week	43.8	1,340,404	1.0
Car & Driver	58.7	1,340,353	8.6
Changing Times	50.5	1,096,672	17.5
Cosmopolitan	198.8	5,756,536	9.4
Discover	50.5	775,732	16.1
Duns Review	7.2	72,504	0.9
Ebony	67.2	1,526,995	8.1
Elks Magazine	10.8	106,080	6.0
Esquire	45.6	788,616	11.4
Essence	52.2	582,510	6.3
Family Circle	194.2	10,824,341	12.5
Family Handyman	14.9	234,277	4.5
Field & Stream	130.1	4,304,230	17.4
O Plus	1.2	6,564	0.4
Forbes	24.1	566,741	1.1
Fortune	5.5	215,686	0.3
Gallery	25.3	162,419	26.5
Games	20.4	194,778	18.6
Gentlemen's Quarterly	7.2	72,126	0.8
Glamour	149.2	3,426,371	7.6
Golf	50.4	867,023	6.7
Golf Digest	24.2	644,993	3.7
Gourmet	31.2	461,748	5.4
Grit	82.5	788,586	29.1
Harpers Bazaar	68.4	1,012,746	7.1
Harpers Magazine	43.2	384,816	22.6
House Beautiful	80.4	1,560,036	8.7
House & Garden	133.0	3,231,088	12.4
Inside Sports	70.8	949,482	21.1
Ladies Home Journal	162.3	7,865,491	16.3
Life	108.0	3,630,032	17.8
Mademoiselle	101.6	1,470,584	7.3
McCall's Magazine	171.4	9,612,510	15.1
McCall's Working Mother	10.8	66,324	1.9
Mechanix Illustrated	107.2	2,821,762	18.4
Metropolitan Home	134.1	1,855,849	20.0
Money	88.0	2,043,047	9.5
Ms.	61.0	605,369	14.8
Nation's Business	9.6	246,780	2.1

Table 18:2 (Continued). 1981 Cigarette advertising revenue

Magazine	Pages of cigarette ads per year	Yearly revenue	Percent of cigarette ads
New West	63.6	580,679	11.4
New York with Cue	107.2	1,664,824	8.8
Newsweek	471.3	30,145,246	15.8
Next	27.6	230,230	20.0
Omni	61.3	946,492	9.6
OUI	82.9	855,908	34.5
Outdoor Life	102.0	2,472,612	16.2
Panorama	8.4	19,740	5.4
Penthouse	143.2	5,615,332	19.5
People	525.6	20,137,840	16.3
Playboy	192.6	11,175,624	15.5
Popular Mechanics	81.2	2,125,746	10.1
Science Monthly	100.5	2,832,427	17.4
Prime Time	2.4	9,517	2.1
Psychology Today	78.0	2,027,993	17.2
Redbook Magazine	183.1	7,850,875	16.1
Road & Track	60.6	1,193,434	9.2
Rolling Stone	100.8	1,703,632	12.3
Saturday Review	36.0	371,603	12.1
Science Digest	16.8	189,678	7.1
Self	79.2	1,129,850	10.7
Signature	22.8	192,504	3.8
Southern Living	85.1	2,298,145	6.8
Sport	80.5	1,524,012	20.5
Sports Afield	65.1	805,820	9.1
Sports Illustrated	432.9	24,611,965	16.9
TV Guide	412.9	30,075,811	12.7
Tennis	13.2	169,130	2.0
Texas Monthly	69.6	654,522	6.1
Time	460.9	40,530,667	17.2
Town & Country	7.2	72,698	0.6
Travel & Leisure	24.0	443,292	2.5
True Story	90.7	1,260,557	16.7
U.S. News & World Report	249.3	10,950,610	14.6
US	95.6	1,425,782	25.2
Vogue	110.0	1,952,276	5.1
Woman's Day	166.6	9,241,187	11.3
Working Woman	58.8	514,296	9.3
World Press Review	1.2	3,251	0.9
Totals	8020.3	308,348,490	

invites us to "Take a puff" of a Salem. A handsome man, accompanied by the caption, "It's Springtime," offers a Barclay to a waiting lady off-camera. Young women flaunt their newly found feminine independence in ads for Virginia Slims and More. The "Man's Man"—the rough and tough cowboy—shouts his supposed virility in Marlboro Country— "where a Man belongs." Benson and Hedges DeLuxe 100 suggests a "touch of class"—with accompanying pictures of caviar, champagne, silver trays and Rolls Royces. Some ads suggest safety by emphasizing low tar and nicotine levels. "ONLY TAREYTON HAS THE BEST FILTER." "MERIT TASTE STANDS ALONE . . . the proven taste alternative to higher tar smoking." And of course, "CARLTON IS LOWEST."

The purposes of advertising

What are these ads really selling? Let's start with the party line. The tobacco industry staunchly maintains that the only purpose of its ads is to compete for existing smokers:

> A large body of professional research shows that advertising does not increase total cigarette consumption. Cigarette advertising is brand advertising. It strengthens brand loyalty or persuades a smoker to switch brands. There are no ads encouraging people to "smoke more cigarettes," only campaigns saying "smoke our cigarettes, not theirs."
> —From an editorial in the *Tobacco Observer*

> Between the time a kid is 18 and 21 he's going to make the basic decision to smoke or not to smoke . . . if he does decide to smoke, we want to get him.
> —L.W. Bruff, Liggett and Myers Vice President, 1962

> Lever Brothers (a soap manufacturer) doesn't advertise its products to convince you to take a bath. Just as the cigarette advertisements are aimed at gaining brand loyalty among smokers, not getting nonsmokers to try a cigarette.
> —Anne Browder, Tobacco Institute

Is it possible that cigarette advertising is not intended to increase the number of smokers? Former advertising executive Emerson Foote gave this answer:

> In recent years, the cigarette industry has been artfully maintaining that cigarette advertising has nothing to do with total sales. Take my word for it, this is complete and utter nonsense. The industry knows it is nonsense. I am always amused by the suggestion that advertising, a function that has been shown to increase consumption of virtually every other product, somehow miraculously fails to work with tobacco products.

George Young, former British Parliamentary Under Secretary of State for Health, agrees:

I totally reject the argument that advertising has no effect on total consumption but simply redistributes consumption between competing brands. I remember an advertising man justifying the amount of money spent on promoting different brands of toothpaste. He assured me that it promoted general interest in dental hygiene, in addition to promoting the particular brands. I believed him, and I believe that the same is true for tobacco.

In Kenya, the British American Tobacco Company is the fourth largest advertiser despite the fact that it has no competition! In the Federal Republic of Germany, one advertising company promotes the pleasures of smoking without mentioning any brand names. Obviously the real aims of cigarette advertising are:

1) to make cigarette smoking appear socially acceptable, and indeed, desirable

2) to distract smokers from their anxieties about the health effects of their habit

3) to lure new smokers, particularly women and children

4) to head off any potential new concerns about smoking.

Reassurance intended

Most cigarette ads suggest that cigarettes are pleasurable, somehow related to fun, good health and youth, and that tobacco is a wholesome product. As noted by Emerson Foote:

> The implied message is "If it is all right to advertise, the product can't be that bad." The converse of this, of which the industry is fully aware, is that if it is not acceptable to advertise, then there must be something wrong with the product.

Right now, smokers are understandably worried and unsure of the legitimacy of their smoking behavior. Cigarette advertising reinforces their behavior by suggesting that lots of good-looking, healthy young people smoke. A 1952 editorial in *The Lancet* noted that smokers themselves convey a message:

> The desire to smoke originates from the advocacy of smoking by the smokers, much of it unconscious. Each time a smoker lights up, he says in effect "I am in favor of the smoking of tobacco." Every smoker is . . . a living advertisement for tobacco.

The insecurity of smokers—and their need for reinforcement—were clearly acknowledged in materials subpoenaed by the FTC from cigarette companies. R.J. Reynolds' 1977 marketing plan for Salem stated explicitly that its ads must associate the cigarette with "emulatable personalities and situational elements that are compatible with the aspirations and lifestyles of contemporary young adults." The same scheme emphasized the use of young adult males who were "masculine, contemporary, confident, self-assured, daring/adventurous, mature." This imagery is obviously intended to make smokers feel they are in good company.

Anything goes

Tobacco promoters appear to consider any form of distortion, deception, cover-up and rule-bending to be fair play. An internal memo dated July 14, 1979 from British American Tobacco (BAT/Brown and Williamson), leaked to the British press, gives some vivid examples of the industry's character. Anticipating an eventual ban on advertising, the memo begins, "The prospects for 1990 are poor," and stresses "the importance of bringing plans to fruition and initiating action *well before* bans or severe restrictions are imposed."

The memo discusses how national bans can be circumvented by "beaming TV and radio advertising into . . . a 'ban' country [although] obviously the political risks of this action must be weighed up and treated with prudence." The memo suggests increasing good will through sponsorship of sporting events and concerts. Companies should also look for "opportunities . . . to find non-tobacco products and other services which can be used to communicate the brand or house name . . ." (An example would be candy cigarettes with nearly identical names and logos of the major cigarette companies.)

The BAT memo advises cigarette companies to make a special effort to court the media, stressing the "candor" of the tobacco industry and its willingness to enter into dialogue:

> Opportunities to establish and nurture friendly relations with media writers and presenters should be sought. Even when a medium is banned, articles and programs sympathetic to the industry can often be published and carefully chosen data should be compiled to take advantage of such opportunities.

In 1982 the Tobacco Institute promised "straight answers to tough questions" in ads directed at journalists through such publications as *Columbia Journalism Review*. The ads pictured Tobacco Institute representative Tom Howard and invited readers to call a toll-free number to get "the other side of the tobacco controversy." But when staff members of the American Council on Science and Health, calling as individuals, tried to reach Mr. Howard by phone, he was *never* available to take calls immediately. When he eventually called back and learned the affiliation of the questioners, he said, "Don't bother calling here again as I am rarely available to receive calls." Thus it appears that the ad was a sham, a public

> **DO CIGARETTE COMPANIES WANT KIDS TO SMOKE? NO.**
>
> There's more than one side to every issue. Including those involving cigarettes.
>
> That's **Tom Howard's** job. Giving straight answers to tough questions about cigarettes. In person or on the phone.
>
> You need the other side.
>
> Call toll-free (800) 424-9876.
>
> **THE TOBACCO INSTITUTE.**

relations effort built on the premise that journalists wouldn't actually call but might remember the ad and think that the cigarette folks—being "open to dialogue"—weren't so bad after all.

Downplaying the health issue

Tobacco industry documents reviewed by the FTC indicate that many cigarette advertising techniques have been aimed at undercutting the health warning. Documents from Brown and Williamson, and one of its advertising agencies, Ted Bates and Company, Inc., focused on ways to reduce public concern about health problems. Based on its research, Bates reported to B&W that many smokers perceive the habit as dirty and dangerous, a practice followed only by "very stupid people." The report concludes:

Smokers have to face the fact that they are illogical, irrational and stupid. People find it hard to go throughout life with such negative presentation and evaluation of self. Tha saviours are the *rationalization* and *repression* that end up and result in a defense mechanism that . . . has its own 'logic,' its own rationale.

The report suggests that good ad copy should de-emphasize the objections to smoking: "Start out from the basic assumption that cigarette smoking is dangerous to your health—try to go around it in an elegant manner but don't try to fight it—it's a losing war."

Other B&W documents show that in marketing Kools, the company was capitalizing upon the "pseudo health image" of mentholated cigarettes. In a 1978 document, B&W admitted that part of Kool's popularity "rides on the connotation that menthol has health overtones," and that the combination of menthol and low tar in Kool Super Longs was even more potent. Actually, mentholated cigarettes tend to have high "tar" and nicotine content.

Other documents reveal that in 1976 B&W had introduced a new brand, Fact, with the assumption that there was less harmful gas in its inhaled smoke. When sales were unsatisfactory, the company considered advertising "more complete health protection through selective gas filtration," but decided that "until the problem of gas becomes public knowledge through government investigation or media coverage, a low-gas benefit will remain of little strategic value." Thus, as the FTC report concluded, despite the market advantage it might have obtained by advertising FACT's new filtration system, "B&W chose not to do so in order to avoid bringing to the attention of the public the hazardous nature of gases in cigarette smoke."

To new markets

Cigarette manufacturers emphasize that smoking is a "pleasure for adults." Tobacco Institute's Horace R. Kornegay wrote to HEW Secretary Califano that, "It has long been the view of the tobacco industry that smoking is an adult custom." But the above-mentioned Bates report indicates otherwise:

> In the young smoker's mind, a cigarette falls into the same category with wine, beer, shaving, wearing a bra (or purposely not wearing one), declaration of independence and striving for self-identity. . . . Thus, an attempt to reach young smokers, starters, should be based, among others, on the following major parameters:
>
> Present the cigarette as one of the few initiations into the adult world.
>
> Present the cigarette as part of the illicit pleasure category of products and activities.

> In your ads create a situation taken from the day-to-day life of the young smoker but in an elegant manner have this situation touch on the basic symbols of the growing-up, maturity process.
>
> To the best of your ability, (considering some legal constraints), relate the cigarette to pot, wine, beer, sex, etc.

Yes, tobacco companies obviously want children to become interested in cigarettes just as cosmetic manufacturers want little girls to learn about lipstick. But the lengths to which tobacco companies go to promote their products is a story in itself. The youth-oriented movie Superman II contains no fewer than 13 separate references to Marlboro cigarettes in background props and ads. Perhaps someday it will be revealed how they got there.

Some tobacco executives even deny pitching specially to women. Edward A. Horrigan, Jr., Chairman of R.J. Reynolds Tobacco Company, attributes the increase in women's smoking not to advertising, but to changing lifestyle: "The woman who is an account executive or creative director emulates the man she knew about 10 years ago who smoked and drank. So she'll have a Scotch or maybe a glass of white wine and light up a cigarette." But Mr. Horrigan also brags about his company's fast-growing sales of MORE cigarettes—and MORE LIGHTS, a reduced-tar offshoot that features beige paper. Women are expected to make up 90 percent of its customers. "Beige has tremendous feminine appeal," says Horrigan.

"Eve" cigarettes are being suggested for the "sophisticated, up-to-date, youthful and attractive woman who seems to have distinct ideas about what she wants." And this has nothing to do with advertising? George Washington Hill would have many a chuckle over that!

Since advertising is legal, should we blame the industry for attempting to remain viable? A company cannot remain long in business without selling to the next generation. Defending the advertising practice of using young, beautiful, well-dressed models, Tobacco Institute's Anne Browder once said, "What do they want us to use for a model, a hobo wearing a torn raincoat and standing in front of a porno store? We have a product to sell."

Heading off concerns

The Tobacco Institute has been trying to influence public opinion by advertising to smokers and nonsmokers alike. In this regard, a confidential report done for the Institute by the Roper Organization ("A Study of Public Attitude Toward Cigarette Smoking and the Tobacco Industry in 1978") and later obtained by the FTC is particularly revealing. Here's a "balance sheet" of the report's conclusions:

ASSETS

1. The overall saliency of the "cigarette issue" is low. Compared to crime, drugs, pollution, and a half-dozen other items, smoking is at the bottom of the list of personal concerns.
2. There is little sentiment for a total ban on cigarettes in public places (but see #3 under Liabilities).
3. There is overwhelming approval of placing notices outside places that restrict cigarette smoking.
4. Few people favor job discrimination based on cigarette smoking.
5. The percentage of smokers in the 17 to 24-year-old age group is up, and the amount smoked per day per young smoker is also up (but see #5 under Liabilities).
6. There is broad support for FTC regulation of "public service" advertising sponsored by non-profit groups like the Cancer Society and Ralph Nader.
7. There is less than majority sentiment in favor of a graduated cigarette tax in which those highest in tar would be taxed the highest.

LIABILITIES

1. More than nine out of every ten Americans believe that smoking is hazardous to a smoker's health.
2. A majority of Americans believes that it is probably hazardous for a nonsmoker to be around people who smoke.
3. There is majority sentiment for separate smoking sections in all public places we asked about.
4. There is majority acceptance of the idea that the cigarette warning label should be made stronger and more specific.
5. The percentage of people who smoke cigarettes is at the lowest level measured in the past ten years.
6. A steadily increasing majority of Americans believes that the tobacco industry knows that the case against cigarettes is true.
7. Favorable attitudes toward the tobacco industry are at their lowest ebb.
8. There is widespread support for anti-smoking education in the schools—and at the very early years.
9. Two-thirds of smokers would like to give up smoking.
10. Nearly half the public thinks that smoking is an addiction.
11. More people say they would vote for rather than against a political candidate who takes a position favoring a ban on smoking in public places.

The report's authors predicted big trouble. Although the tobacco industry had apparently survived several decades of bad health news for cigarette smokers, the anti-smoking forces' latest tack on the passive smoking issue is another matter. What the smoker does to himself may be his business, but what the smoker does to the nonsmoker is quite a different matter. The anti-smoking forces have not yet convinced as many people that smoking harms the health of the nonsmoker as they have convinced people that smoking harms the health of the smoker. But this study shows that they are well on the way. . . . This we see as the most dangerous development to the viability of the tobacco industry that has yet occurred.

The report urged the industry to "develop and widely publicize clear-cut, credible, medical evidence that passive smoking is not harmful to the nonsmoker's health." That is what the industry has tried to do. *Indeed, the fear that concern about smoking might spill out beyond the smokers themselves put the industry on red alert.* Hundreds of millions of dollars have been spent to encourage smokers and nonsmokers to "co-exist," and to downplay the claims that passive smoking is hazardous.

Why cigarette advertising should be banned

It is clear that cigarette advertising is socially harmful. Some 50 countries now have taken legislative action or entered into voluntary agreements imposing restrictions on advertising of tobacco products. Of these, 15 ban all advertising. This group includes socialist countries which prohibit all advertising.

Singapore in 1970 prohibited all advertising of cigarettes, cigars or any other form of tobacco and applied a very broad definition of "advertising," including notices, circulars, pamphlets, brochures, programs, price lists, labels, posters, placards and billboards. In 1973, Norway passed a law simply stating that "advertising of tobacco products is prohibited," but adding that if a safe cigarette ever came along, the government would reconsider this issue. In 1976, Finland enacted a total ban, except for ads in foreign print publications, stating, "Advertising for tobacco, tobacco products and imitations and smoker's accessories and other sales-promotion activity directed at the consumer, as well as their association with advertising for other products or services, or other sales promotion activity shall be prohibited." In 1980, Bulgaria, a major tobacco-producing country, banned all TV, radio, movie and press ads, as well as promotion on posters, signs and other such places.

While the United States has no cigarette advertising on radio or television, it faces more than a billion dollars' worth of newspaper, magazine and billboard ads each year as well as ads for smokeless tobacco products. What is the appropriate policy for our free-enterprise society? Given the tobacco industry's long history of irresponsible advertising, a voluntary approach obviously won't work.

The Cigarettes Advertising Code, introduced by the tobacco industry and in effect between 1964 and 1970, had little effect on the nature of the ads. (Remember, for example, how American Tobacco pledged not to appeal to young people but subsequently advertised that, "Luckies separate the men from the boys, but not from the girls.") Yet the industry's line remains that "there are reasonable and necessary limits on any advertising. Sensible laws protecting consumers against false or misleading claims are clearly appropriate. What is inappropriate are unwarranted restrictions on advertising."

That's generally true, but cigarettes and their advertisements present a unique situation. Not only are cigarettes harmful—but the tobacco industry has never *admitted* that they are harmful. Alcoholic beverage ads often warn against excess consumption and the high risks of driving while intoxicated. But cigarette advertisers cannot claim that there are safe and unsafe ways to use their products. So they pretend that tobacco is as harmless as popcorn.

Some countries have restricted ads in magazines read by children, eliminated ads for high-tar/nicotine cigarettes, forbidden the use of human models in ads, ordered more prominent warnings, and banned billboard ads or tobacco sponsorship of events. But the most effective and sensible policy for the United States would be *an immediate total ban on cigarette advertising.*

Obviously, the tobacco industry would object violently. Would such a ban be constitutional? The U.S. Supreme Court (where the matter would inevitably end up) would certainly discern that the health effects of smoking cigarettes are overwhelmingly negative—and would probably conclude that the gains from banning the ads would far outweigh any loss of corporate freedom.

An advertising ban would probably not cause a dramatic reduction of smoking in the generation of smokers already hooked. But in the long run, it would cut sales by communicating clearly that cigarettes are no longer socially acceptable. Equally important, a ban would leave magazines and newspapers free to discourage cigarette smoking without fear of economic reprisal. Tobacco companies, which have already diversified, would of course be free to apply their marketing skills to more worthwhile industries.

What can you do

- Write to the editors of your favorite magazines and newspapers, asking them to consider discontinuing cigarette advertising. Point out that it simply is not consistent with their presumed interest in the welfare of their readers to run ads for a commodity known to be hazardous to human health. (Expect the standard bedbug letter the first time you write; it will tell you that they are advertising because it is a legal product, and they think their own readers have enough sense to decide for themselves. Write back noting the amount of income they derive from the ads, asking if they could begin to replace the cigarette ad copy with more legitimate, consumer-oriented types.)

- Write to your Congressional representatives, encouraging them to introduce or support legislation to ban all cigarette advertising. Emphasize that your interest in making this recommendation is based on your belief that the elimination of such advertising will tarnish the glow of cigarette acceptance, foster the idea that smoking is unacceptable, and encourage smokers to discontinue their habit.
- Write to the Federal Trade Commission, indicating your support for a ban on cigarette advertising.

19
Nonsmokers Should be Protected

Most people are sensitive to tobacco smoke to some degree.

When nonsmoker Richard Lent boarded an Eastern Airlines flight from Washington, D.C., to New York City and found no seats available in the nonsmoking section, he took seat 27B—five rows into the designated smoking section. Unwilling to breathe the exhaust from fellow passengers' cigarettes, Lent insisted that the nonsmoking section be extended to include his seat. In accordance with Civil Aeronautics Board (CAB) regulations, the stewardess moved the "No smoking" sign back to Row 27. When smokers who had lost their smoking status lit their cigarettes anyway, a bitter argument ensued, with passengers shouting and standing in the aisles. Unable to quell the altercation, the pilot made an emergency landing in Baltimore, delaying the flight's arrival in New York by three hours.

Incidents like the above reflect a complex and unsettled issue: when smokers and nonsmokers must share the same airspace, whose "rights" should prevail? Twenty years ago, this question would never have been raised. Smokers constituted a majority, at least among men, and smoke-filled areas were accepted as a normal part of life.

In 1964, when the Surgeon General's report established the link between smoking and disease in the public's consciousness, more than half of American men and some 30 percent of American women smoked cigarettes. Today only 37 percent of adult men and 29 percent of adult women are smokers. Now a clear majority, nonsmokers are saying with increasing frequency, "Yes, I do mind if you smoke," and their protests are being acknowledged.

In 1973, for example, the CAB, which regulates U.S. airlines, mandated separate smoking and nonsmoking sections in airplanes and required that all nonsmokers be guaranteed a space in the no-smoking section if they so requested. As this book goes to press, the CAB is considering a *total* smoking ban on flights of less than two hours' duration.

People have also become less tolerant of smoking in the workplace. Many companies have voluntarily banned or restricted smoking. San Francisco voters recently approved an ordinance requiring employers to draft policies accommodating the preferences of nonsmokers.

Many communities have enacted legislation restricting smoking in public places. Restaurants, hotels and even car rental agencies have become increasingly sensitive to the preference of the nonsmoking majority, offering no-smoking tables, rooms and cars.

Fueling the nonsmokers' rights movement have been reports that drifting cigarette smoke presents not merely an annoyance but also a health hazard to nonsmokers. Unfortunately, there has been some exaggeration on both sides about the health effects of "second-hand" smoke. Some activists have insinuated that spending time in the vicinity of a burning cigarette is nearly as dangerous as smoking one. The tobacco industry, on the other hand, has pointed to flaws in some of the more well-publicized studies on passive smoking, falsely implying that all of the evidence on this subject is scientifically unsound.

Everybody's business

Smokers often defend their habit with the contention that cigarette smoking is a private matter and that it is nobody else's business if they choose to expose their own lungs to the harmful effects of tobacco smoke. But this is not true. An average cigarette burns approximately 12 minutes. An average cigarette smoker inhales for only 24 seconds of these 12 minutes. As the American Lung Association points out, two-thirds of the smoke from a burning cigarette is released into the air that *others* must breathe.

It is often impossible to filter this smoke out of the air, especially in enclosed spaces. A study conducted by the American Academy of Allergy showed that even with the most modern equipment, cessation of smoking was necessary to reduce carbon monoxide, a potentially harmful constituent of tobacco smoke, to safe levels. Another study, carried out on a one-hour Boeing 707 flight found that the ventilation system on board did little to protect nonsmokers from the tobacco smoke. The air in the nonsmoking section contained just as many tobacco particulates as did the air in the smoking section. And measurements taken at a typical college campus party demonstrated a particulate level 40 times higher than the U.S. air quality standard!

Clearly, unless the smoker indulges in total isolation, his smoking *is* somebody else's business.

Tobacco smoke contains over a thousand substances. Some, including tar, carbon monoxide and nicotine, are present in *higher* concentrations in sidestream smoke (the smoke released directly into the air between puffs) than in the mainstream smoke which is drawn into the smoker's lungs.

More than 40 constituents of tobacco smoke are known carcinogens, and these too are often present in much higher concentrations in the sidestream smoke. Nitrosamines, which may be hazardous in concentrations as low as 1 part per million are present in sidestream smoke at concentrations 50 times higher than in mainstream smoke. Thus a nonsmoker spending one hour in a smoky room could inhale the same amount of cancer-causing nitrosamines as a person smoking 15 cigarettes.

Quinoline, a chemical which can cause liver cancer in rats, is found in sidestream smoke in 11 times the concentration of inhaled smoke. Benzpyrene, another potent carcinogen, is present at three times the concentration found in mainstream smoke, and just one burning cigarette can release substantial quantities into the air.

Other chemicals contained in relatively high concentrations in sidestream smoke include cadmium, ammonia, nitrogen dioxide, formaldehyde, hydrogen cyanide, arsenic and hydrogen sulfide.

Judging from the variety of potentially harmful substances known to be present in tobacco smoke (many, no doubt, are as yet unidentified), it would appear that nonsmokers who live in, work in, or otherwise occupy smoke-filled spaces have ample cause for concern over their health. But the current epidemiological and experimental evidence deserves careful interpretation.

Immediate effects

Apparently, most people are sensitive to tobacco smoke to some degree. In a survey of more than 1,000 Toronto residents, almost 90 percent said they were affected by cigarette smoke. A survey of undergraduates at a New Hampshire college also found that a sizable percentage of nonsmokers reported adverse physical and psychological reactions to cigarette smoke. The most frequent symptom reported as a result of exposure to tobacco smoke is eye irritation, evidenced by tearing, burning and increased blinking. Coughing, nasal discharge, stuffiness, headaches and throat irritation are also common.

Laboratory experiments have shown that exposure to ambient carbon monoxide at levels similar to those in tobacco smoke can temporarily increase blood pressure and heart rate and decrease the oxygen-carrying ability of the blood.

The immediate effects of exposure to cigarette smoke can be more serious or pronounced for certain sensitive individuals. In one study, persons with coronary heart disease exposed to the smoke of 15 cigarettes over a 2-hour period displayed an increased resting heart rate, increased blood pressure and decreased oxygen-carrying capacity of the blood. They also became more susceptible to chest pain (angina pectoris) during exercise.

Asthmatics may develop respiratory distress due to cigarette smoke exposure. Allergic individuals are also more likely to suffer adverse reactions such as runny nose and wheezing. Contact lens wearers are more prone to experience eye irritation.

The immediate effects of tobacco smoke exposure are not limited to physical reactions. Adverse psychological reactions are seen as well. While annoyance at being forced to inhale smoke-laden air is certainly not life-threatening, it is a real source of discomfort to nonsmokers. Many individuals, especially those with chronic lung disease, are worried that involuntary inhalation of tobacco smoke may harm them.

Long-term effects: cancer?

In 1981 the headline "Cancer Study Reports Higher Risk for Wives of Smoking Husbands" blazed across the front-page of *The New York Times* and numerous other publications. The headlines referred to a 14-year epidemiological study conducted at the National Cancer Center Research Institute in Tokyo by Dr. Takeshi Hirayama. The results of Hirayama's study were indeed astounding and unprecedented, indicating that wives of Japanese men who smoked were up to twice as likely to die of lung cancer as were wives of nonsmokers. In fact, the risk to "passively smoking" women in Hirayama's study was one-third to one-half that of women who themselves smoked!

Hirayama asserted that his results "appear to explain the long-standing riddle of why many women develop lung cancer although they themselves are nonsmokers." Other prominent scientists, criticizing Hirayama's study for improper experimental design and analysis, were not so readily convinced.

One major problem was the fact that the age-adjusted lung cancer mortality rates for the entire study population and all of its subcategories were higher than those in Japan as a whole. Even nonsmoking women with nonsmoking husbands had higher lung cancer death rates than did smokers, passive smokers and nonsmokers in the rest of Japan, indicating that another factor may have been responsible for the lung cancers observed by Hirayama.

Another problem was the fact that a majority of the lung cancers which occurred in his study population were adenocarcinomas, a type of lung cancer not generally associated with cigarette smoking. Failure to standardize for the womens' ages and possible errors or undocumented changes in smoking status classification cast more doubts on the results, as did allegations that Hirayama had made a serious mathematical error in his calculation. (Hirayama was later vindicated on the mathematical error charge.)

Finally, the sheer magnitude of the effect of passive smoking on the women in Hirayama's study led many epidemiologists to question the plausibility of his results. The two-fold increase in lung cancer risk which Hirayama reported for wives of heavy smokers is close to that observed for women *actively* smoking *five* cigarettes per day. Since the men who were heavy smokers smoked, on the average, only about eight cigarettes per day at home, this would suggest that exposure to the smoke of a cigarette is nearly equivalent to actively smoking it. This seems highly unlikely, especially in view of autopsy studies which have shown that among people who died of causes other than lung cancer, hardly any nonsmokers displayed evidence of precancerous lesions in the lungs, while a significant number of smokers did. If exposure to cigarette smoke were an important factor in producing lung cancer, one would expect to see these lesions in a substantial number of nonsmokers as well.

Unfortunately, both the pro-tobacco forces and uncritical anti-smoking zealots have continued to cite Hirayama's study as evidence for their cause. Nonsmokers' rights activists have touted the flawed study as "proof" that second-hand smoke is endangering their lives, while the tobacco industry has not hesitated to publicize it as a prime example of the shoddy and insubstantial evidence used to (wrongfully) incriminate tobacco.

Six months after Hirayama had dropped his bombshell, Lawrence Garfinkel of the American Cancer Society published the findings of another major epidemiological study on the role of second-hand cigarette smoke in causing lung cancer. Appearing in the *Journal of the National Cancer Institute,* the study's results contradicted those of Hirayama.

Among 375,000 nonsmoking women and 148,000 nonsmoking men, Garfinkel detected no consistent increase in lung cancer deaths over the 17½-year study period. If "passive smoking" were an important factor in the development of lung cancer, an increase in lung cancer deaths over time would be expected. But Garfinkel found no significant differences in lung cancer rates between women whose husbands smoked and those whose husbands did not smoke.

A recent study of Greek women found that women who developed lung cancer were more likely to have husbands who smoked than were women from the same area who did not have lung cancer. The number of subjects in this study was small, however, and the results are not generalizable to other groups of women.

A fourth study of nearly 4,000 Louisiana residents indicated a relationship between exposure to a spouse's cigarette smoking and the development of lung cancer.

Thus, at present, there is suggestive evidence of an association between lung cancer and passive smoking, but additional research is needed to confirm or refute this.

Long-term effects: respiratory impairment

By using special devices which measure a person's ability to inhale and exhale air, scientists can ascertain how well the small bronchial tubes and air sacs in the lung are functioning. Physiologist James White and physician Herman Froeb administered such tests to 2,100 middle-aged workers. They found that nonsmokers who had worked in a smoky environment for more than 20 years showed slight but significant changes in lung function tests in comparison to nonsmokers who neither lived nor worked in a smoky environment. Men and women chronically exposed to tobacco smoke had test scores similar to light smokers on lung function tests. Another study of children whose parents smoked also demonstrated impaired lung function.

Second-hand smoke—in the womb

Many dangerous substances which the mother inhales into her own lungs pass into her blood stream and affect the delicate tissues of the baby just as do drugs that the mother might ingest. Smoking just one or two cigarettes significantly slows breathing movements and increases the heart rate of the fetus, indicating clearly that the immediate effects of smoking are not limited to the mother.

Pregnant women who smoke are more likely to miscarry or have still-born babies than are nonsmoking women. They are also more likely to deliver prematurely. The Surgeon General has estimated that 14 percent of premature deliveries, with their accompanying risks to the infant, are caused by maternal smoking.

Babies born to smokers are generally less alert, less vigorous, and smaller than babies born to nonsmokers, weighing an average of 6 ounces less. Low birth weight is associated with increased probability of infant death. Even more frightening, there appears to be an association between maternal smoking and Sudden Infant Death Syndrome (SIDS). One study of 20,000 infants found that the risk of a baby dying from SIDS was over 4 times greater if its mother smoked during pregnancy than if she did not.

If the mother smokes after the child is born, the child runs an increased risk of respiratory infections such as bronchitis or pneumonia during the first year of life. If the father spends an appreciable amount of time in the presence of the infant, his smoking may also adversely affect its health. Several studies indicate that older children are more likely to suffer from respiratory illness if they live in a household where one or both parents smoke.

A sensitive issue

Exposure to second-hand smoke is surely an annoyance and an irritant to most nonsmokers and a threat to the health and life of the unborn baby. Evidence also indicates that chronic exposure to cigarette smoke may lead to subtle, physiological changes in the nonsmoker. While many studies have suggested that more serious health effects may be attributed to passive smoking, the weight of the evidence does not yet warrant the inclusion of smokers in the same category as axe murderers, typhoid carriers and other such dangerous characters.

In fact, the misplaced hostility which has sometimes been directed toward smokers rather than their offensive habit, has probably done a great deal to undermine nonsmoker's rights efforts.

Smokers, understandably, may become quite defensive and recalcitrant toward what they perceive as spiteful efforts to curtail their civil liberties. The tobacco industry aggravates these hostilities by depicting nonsmokers' rights activists as a bunch of proselytizing do-gooders who will be satisfied with nothing less than a total prohibition of tobacco.

Actually the tobacco industry is quite frightened by the current surge of activism in the field of nonsmokers' rights. A 1978 issue of *The Tobacco Observer* called it "the most dangerous development to the viability of the tobacco industry that has yet occurred."

Nonsmokers are a group over which tobacco has no hold, so there is no irrational habit to override their rational concerns about health and comfort. Increasingly, nonsmokers are banding together to get state and local governments to restrict or prohibit public smoking. Well over two-thirds of the 50 states and 36 countries have enacted laws to restrict smoking in public areas. Many cities have anti-smoking laws and many businesses have voluntarily instituted nonsmoking rules.

The tobacco industry is of course worried that trends toward curtailing public smoking will not only reduce the number of cigarettes consumed, but will eventually make smoking a socially unacceptable activity. This thought is intolerable to an industry spending millions of dollars each year to convince young people that smoking is a mark of sophistication and social success. As smokers become more of an unwelcome minority, it will become increasingly difficult to convince potential smokers that lighting up is socially advantageous or smart.

Previous experience with public spitters suggests that such a mechanism for reducing the number of smokers might prove successful. During the late 19th century, it was quite common for gentlemen to expectorate tobacco and phlegm onto the nearest available surface. While there were certainly many who must have objected, it took the new medical knowledge that the health of "non-spitters" might be harmed by germs present in the spit to bring about unified action against public spitting. "Do Not Spit" signs were posted, the Tuberculosis Association organized a war on spitting, and finally, anti-spitting legislation was passed in many cities and states.

No doubt many spitters objected strenuously to this encroachment on their freedom, but spitting soon went from being a socially acceptable custom to one which drew general public contempt and disgust. This social pressure was apparently enough to cause the spitting habit to disappear almost completely, in private as well as in public. Nowadays, one would have to look long and hard to find a spitter, or even a spittoon. Perhaps the same will be said about smokers and ashtrays in the not too distant future!

Recommendations for action

Until there are more definitive data on the gravity of the health hazard posed by exposure to second-hand smoke, perhaps the most prudent way to approach the issue is to assume that cigarette smoke annoys most nonsmokers, definitely harms some nonsmokers, and may pose a health hazard to the rest.

Certainly, annoyance alone is a valid basis for demanding restriction of smoking in public places. If someone were to squirt you with a water pistol every time you rode the bus or dined in a restaurant, it might not endanger your health, but it would certainly annoy you; and few would question your right to take action to halt this annoyance. Nonsmokers have the same right to take action to stop the annoyance of breathing smoke-filled air or suffering from itchy, tearing eyes caused by exposure to cigarette smoke.

While efforts to protect the rights of nonsmokers have produced some very gratifying results, there still remain frontiers where further action is needed. These should be the concern of all who are interested in the right of everyone to live in a comfortable, healthful environment.

National anti-smoking laws like those enacted in many other countries should be one of the ultimate goals of nonsmoker's rights activists. Finland's anti-smoking regulations provide a model for national control of smoking in public places. Instead of listing certain areas where smoking is prohibited, it prohibits smoking in *all* public places, then specifies certain exceptions where smoking *is* allowed. The law also specifies that at all events attended by children, a "No Smoking" sign must be visible to every person present. This legal insinuation that nonsmoking is the norm (which it is) and that smoking is a deviation from the norm (which it is) may prove to be a powerful persuader.

In the absence of national legislation, all states which do not have antismoking laws should enact them. Municipalities may also find it useful to implement their own laws.

Certainly, health organizations, agencies or clinics should set the example by prohibiting smoking in their offices, conventions, waiting rooms, etc. Hospitals should prohibit smoking by all on-duty staff and by patients and visitors in all wards and semi-private rooms except those exclusively designated for occupancy by smokers.

All businesses should keep in mind the ill effects that working in a smoke-laden environment can have on the morale and productivity of nonsmokers. Nonsmoking areas are appreciated by nonsmokers and are usually accepted with minimal complaint by smokers.

Any industry that uses or produces substances known to act synergistically with tobacco smoke in causing disease should immediately ban all smoking in areas where such chemicals are in use, to protect the health of all workers.

Restaurants, supermarkets, theaters, banks, and other establishments which have public traffic should ban smoking or provide nonsmoking areas whether legally required to do so or not. Nonsmokers are the majority and will certainly react favorably to a demonstration of concern for their comfort and welfare. It seems unlikely that smokers (the minority) would stop patronizing establishments that restrict smoking.

What you can do

If you are concerned about the rights of nonsmokers to live in a smoke-free environment, the following suggestions may be of interest.

- If you don't smoke, don't be unnecessarily militant toward smokers. Remember, it is their smoking and its devastating health and economic effects which are objectionable, not them personally. Stress that you simply want to protect yourself, not dictate their behavior.
- Make it your responsibility to see that no-smoking rules are enforced, by politely pointing out the regulations to violators, and/or management.
- Do speak up if someone's cigarette smoke is bothering you, even if there's not a "No Smoking" sign present. Most smokers will be considerate enough to put out their cigarette if it bothers you.
- Get "Thank You for Not Smoking" signs for you home and office, available through your local chapter of the American Cancer Society. Place them on your outside doors as well.
- Insist on your right to a no-smoking seat in airplanes, trains and busses. If drifting smoke bothers you, be sure to complain to the company president and appropriate government regulatory agencies.
- Complain to the person in charge if you are exposed to drifting smoke at a bank, a post office, an airport waiting area, or anywhere else where you think the exposure is unfair.

- If you are trying to convince a supermarket or store to implement no-smoking rules, first ask the management. If it is not receptive, shop at a competing store which does have no-smoking rules. Then send your receipts with a letter of explanation to the offending store.
- If you work in a smoky environment, ask your boss to provide nonsmoking areas for all nonsmokers.
- Loudly, but politely, request a seat in the nonsmoking section of a restaurant, even if you know they don't have one. The restaurant owner and fellow patrons may eventually get the message.
- Keep abreast of current proposals for anti-smoking legislation in your city, county or state.
- Join and support national and local nonsmokers' rights groups such as Action on Smoking and Health (ASH) or Group to Alleviate Smoking Pollution (GASP).
- If you do smoke and are pregnant, planning to become pregnant, or are already a parent, stop smoking *now!* Your child's life and health are much too precious to risk unnecessarily.

Smokers, do you want to support an industry that doesn't care if it kills you or someone you love?

20
The Smokescreen Must Be Lifted

"Men do not usually die, they kill themselves."—Michel de Montaigne
(1533-1592)

Everyone knows that smoking is dangerous to human health. Well, almost everyone. Gallup and Roper polls taken in the late 1970s concluded that about 10 percent of American adults did not believe that smoking is hazardous, and that smokers were more apt to think this than nonsmokers. But many who knew *generally* of smoking's danger were unclear about the *specific* risks involved.

For example, 30 percent of Americans did not know that a 30-year-old man reduces his life expectancy by smoking a pack a day or that smoking less than one pack a day is still dangerous. Forty-three percent were unaware that smoking is a cause of heart disease. And some 80 percent did not realize that smoking causes most cases of chronic bronchitis and emphysema.

It seems logical to assume that if smokers understood fully—and could personalize—the risks assumed by lighting up, they would be less likely to continue. Similarly, if nonsmokers could comprehend the magnitude of the health problems related to cigarettes and their enormous contribution to our country's medical bills, they might become agitated enough to take action. But since the tobacco industry knows that its best customer is an uninformed one, it has been doing everything in its power to maintain a smokescreen around the facts.

Freedom of choice?

One of the industry's standard lines of defense is to stress "freedom of choice" in smoking. But are smokers really free? When *Chicago Tribune* columnist Bob Greene invited his smoking readers to explain why they persisted in using cigarettes, he expected a flood of angry letters. Instead he got replies like these:

> Don't refer to what we do as smoking—refer to it as "nicotine addiction." Don't refer to "quitting"—refer to "nicotine detoxification." Nobody really likes smoking all that much. It's the nicotine surges that we get hooked on. I am trying to quit. Detoxification is truly the most difficult thing I've ever experienced—and I haven't had a particularly uneventful life. If I give in to the addiction, I know I'll die with a cigarette in my hand.
>
> God bless those who have found the strength to quit smoking—but don't condemn those of us who have yet to find an effective escape. Nonsmokers . . . have no idea of the depth of the smoking habit. What simply annoys you is actually killing us.

If smokers could simply quit whenever they wanted, there might be something to back up the "free-to-choose" line of the industry. But physical and psychological dependence represents *involuntary* continuation. Dr. Robert Dupont, former director of the National Institute on Drug Abuse, estimates that only 10 to 15 percent of the people alive today who ever used heroin are still addicted, whereas more than 66 percent of those still living who ever smoked cigarettes still smoke daily. Only 2 percent of smokers consume cigarettes on an occasional basis, and the average smoker consumes 30 cigarettes per day. At the very least, young Americans should be clearly informed that taking that first drag at an early age may be the first step toward irreversible addiction.

Cigarette additives can be dangerous

Your favorite packaged foods undoubtedly tell all on their labels. You know what you are getting, even if that list of unpronounceable chemicals makes you a bit nervous. Not so with cigarettes, described by the National Cancer Institute as a "unique chemical factory, generating more than 4,000 known compounds." Some of the chemicals in smoke are the result of naturally occurring compounds burned at high temperatures, while others result from additives. The first Camel in 1913 was laced with chocolate. Since then, the industry has been adding agents to alter the taste, to moisten, and to slow the burning of the cigarette. Manufacturers say that they can choose from more than 1,400 tobacco additives.

When the demand for filters and other "light" brands picked up in the 1970s, the industry realized that if tar content were lowered, it would be necessary to add chemicals so that the cigarette would "satisfy." As one tobacco chemist told the *Wall Street Journal,* "If there weren't any flavorants in any of these low-tar and low-nicotine cigarettes, you would taste nothing. . . . It would be like smoking a piece of paper."

In 1979, U.S. Surgeon General Dr. Julius Richmond became concerned about these added chemicals, noting that even those cleared for human consumption had never been shown to be safe when *burned*. Public Health Law 95-626 requires the Secretary of Health and Human Services to study "the relative risks associated with smoking cigarettes . . . containing any substances commonly added and report this information to the Congress." But when Dr. Richmond asked the six major tobacco companies for a list of cigarette additives, he was told that they were "trade secrets."

What *is* known about cigarette additives is not reassuring. The most common flavorings seem to be cocoa, licorice, and fruit juices. Not much is known about the effects of licorice and fruit, but the National Cancer Institute reported in 1977 that condensed tar from cocoa-flavored cigarettes caused more tumors in mice than did the tar from unflavored ones. "Based on our results, cocoa should probably not be added to cigarettes," says Thomas Owen, the head of NCI's Smoking and Health Program.

To keep cigarettes fresh, humectant chemicals are added, the major ones being glycerol and glycols. Forty million pounds of them are used each year to keep tobacco moist. When burned, glycerol produces acrolein, a chemical which interferes with the normal clearing of the lungs. And according to the 1970 Surgeon General's Report, "glycols are suspected to influence the smoker's risk of bladder cancer."

There is also coumarin, a chemical that gives cigarette smoke a sweet aroma and a taste like that of fresh-cut hay. After tests in the 1950s showed that it can poison the liver and other organs, the FDA removed coumarin from its Generally Recognized as Safe (GRAS) list of food and drug additives. But neither the FDA nor any other agency has legal jurisdiction over cigarette additives.

So as health-conscious cigarette smokers switch to low-tar brands, they have something new to worry about: untested, unapproved chemicals. Only the tobacco companies and some of their suppliers of flavoring know what these chemicals are and what compounds they produce when burned.

Menthol is another widely promoted cigarette additive. It is an anesthetic. The reason why some smokers find mentholated cigarettes "cooler" is that their throats are literally left numb.

Warning labels

The staff of the FTC has recommended to Congress that more action be taken to educate consumers about the risks of smoking. The agency wants warning labels to be made more specific and conspicuous on packages and in advertising. At this writing, cigarette labeling and education bills are being considered in both houses of Congress. The House bill would require a series of rotating warnings on cigarette packages and in advertising. Each of the following warnings would appear an equal number of times throughout the year:

> WARNING: Cigarette Smoking
> -causes LUNG CANCER and EMPHYSEMA
> -is a major cause of HEART DISEASE
> -is ADDICTIVE and may result in DEATH
>
> WARNING: Cigarette Smoking
> by Pregnant Women May Result in
> MISCARRIAGE, PREMATURE BIRTHS or
> BIRTH WEIGHT DEFICIENCIES
>
> SMOKERS: No Matter How Long You Have Smoked
> *QUITTING NOW* Greatly Reduces the Risk to Your Health

The Senate bill would mandate a single warning that is more specific than the current one:

> WARNING: Cigarette Smoking Causes CANCER,
> EMPHYSEMA and HEART DISEASE; may complicate
> PREGNANCY; and is ADDICTIVE

Backers of these bills range from liberal Democratic Representative Henry Waxman of California to conservative Republican Senator Orrin Hatch of Utah. Also supporting the labeling proposals are such organizations as the American Lung Association, the American Cancer Society, the American Heart Association and the American Dental Association.

Naturally, the cigarette companies do not want stronger warning labels. They may be afraid that the stronger messages might be a step toward banning advertising, but they certainly know that scary language will put off some smokers.

Where has the Reagan Administration been on this important issue? Sitting on the fence. Budget Director David Stockman stated that the Administration "takes no position" because "little is known ... about the relative efficacy of the many alternative labeling schemes being proposed."

Increasing public awareness

Education—including constant reminders to both smokers and non-smokers—is an integral step in dealing with the tragedy of cigarettes. School programs starting in the elementary grades are a must. Voluntary health organizations have a major role to play in distributing educational literature and commentary. FTC Commissioner Michael Pertschuk suggests that cigarette victims and their survivors may prove to be the most effective spokespersons:

Sports heroes and other national figures, especially those who have themselves suffered or had close friends or relatives suffer from smoking-caused diseases, can be enormously effective.... Recently in the United States, groups of mothers whose children have been injured in automobile accidents and groups of high school students themselves whose friends have been killed or maimed... have become powerful spokespersons for legislation to control the sale of alcoholic beverages to young people.

The media smokescreen

Why don't we hear more now about the dangers of smoking and the dishonesty and deviousness of the tobacco industry? Many editors claim that the topic of smoking is "old news" and "boring," and that their "readers aren't interested in rehashing bad news." While there may be some truth to that assertion, it overlooks the fact that there *are* new angles each week to the story of the smoking gun. For example, readers might well find it interesting to learn about: 1) what went on behind the scenes before R.J. Reynolds launched its 1984 "open dialogue" campaign; 2) the latest study showing comparative risks of low-tar versus high-tar cigarettes; 3) recent cases where smokers or their survivors sued tobacco companies; 4) tobacco industry strategy to get children interested in smoking; and 5) comparative death rates in their own communities.

The real reason for minimal news coverage is that recipients of cigarette advertising dollars are reluctant to offend their advertisers. Even radio and television stations share this concern! Although cigarettes themselves cannot be advertised on radio or television, tobacco company subsidiaries heavily advertise other products (such as Seven-Up, owned by Philip Morris) and sponsor sporting events on these media. This situation is far from hopeless. Tobacco companies and subsidiaries must keep advertising to attract customers. If enough media leaders could be persuaded to end the "conspiracy of silence," none would be singled out for "punishment." It is obvious that if tobacco advertising were legally banned or severely curtailed, coverage of the cigarette tragedy would increase dramatically.

Another significant reason that cigarettes have maintained their grip on America is the fact that individuals and associations who seem likely sources of resistance to smoking have failed to speak up. While they may not support the promotion of smoking, they have chosen to tolerate it.

The women's movement

The women's movement, as represented by the National Organization for Women, *Ms. Magazine* and other such associations, are committed to improving the quality of women's lives. The movement has been active in reducing job discrimination, fostering equal opportunity in advanced education, reducing the gap in compensation between male and female workers, proclaiming women's right to control of their own bodies by encouraging availability of birth control services and abortion, and developing novel ways of effectively combining careers with family life.

It thus seems strange that there is virtual silence on the subject of women's smoking in feminist circles. The feminist health "bible," *Our Bodies Ourselves*, makes only passing reference to the subject of smoking. When asked about this, Judy Norsigian, a member of the Boston Women's Health Collective which produced the original book, explained that they had intended to include a chapter on smoking, alcohol and drugs, but "there was not sufficient room in the book, and we did not have the resources to do the research." The National Organization for Women, which has taken a very active role in many women's health issues, has not commented on smoking. Its 40-page submission to the 1979 Kennedy hearings on women's health contains not a single reference to the problem. The National Women's Health Network, which represents a thousand American women's health organizations, has "no formal position on smoking." The San Francisco Women's Health Collective, an organization which describes itself as devoted to "women's health education," does not address the smoking issue because "it [is] not a priority in terms of health education."

Even the magazines of "liberated" women stand mute on this subject. *Ms.*, which claims to "serve women as people, not roles," will not accept advertising considered offensive to women. *Ms.* turned down ads for vaginal deodorants but permits cigarette advertising. It did reject Virginia Slims ads headlined "You've come a long way, baby," but it was "baby" that troubled them, not the cigarettes themselves. Since *Ms.* began publication some ten years ago, it has *never* carried an article on smoking and health.

Why the silence? In the case of the magazines, it is clear that advertising revenues play a role. But there may be more to it than that. Feminist groups traditionally focus on problems which they feel are unique to women, and perhaps they unconsciously decided that the cigarette problem affected everyone. Moreover, the women's movement has always emphasized what others (particularly men) are doing to women as opposed to what women are doing to themselves. To address the growing calamity of cigarette smoking in women, feminist leaders must acknowledge that the problem is largely self-induced. But with lung cancer now replacing breast cancer as the leading cause of cancer death in women, and with cigarettes being the major controllable threat to the health of unborn children, shouldn't priorities be re-examined? Aren't protests and boycotts in order against the companies marketing "female" cigarettes like More, Eve and Virginia Slims?

Two recent events suggest that these may be coming. First, the President's Advisory Committee for women, "Voices for Women, 1980," became the first women's group to acknowledge the hazards of cigarettes to women. Referring to cigarette smoking as the "leading controllable cause of rising morbidity in women," they noted that "smoking may well prove to be the major health problem facing women in the 1980s," Second, British feminist Bobbie Jacobson met the issue head on in her book, *The Lady Killers: Why Smoking is a Feminist Issue*. Her intention in writing it was "to help destroy the myth that lung cancer and heart disease are a male preserve," and to express her displeasure over the "lack of priority that women's organizations—feminist or otherwise—give to the problem."

Religious institutions

While religious organizations can differ widely in their codes of theology, all believe that human life is precious. All would probably agree with the premise that it is a "sin," however defined, to take one's own life. Many have taken formal stances on potentially life-threatening aspects of our lifestyle. For example, the National Council of the Churches of Christ in the USA has issued official policy statements on drug and alcohol abuse. But the Council—and most of the country's formal religious sects—have been silent about cigarettes and health. (A vivid exception here is the Seventh-day Adventist Church, which issues substantial quantities of literature and offers smoking cessation programs throughout the country.)

Why would clergymen who believe that shooting oneself in the head is contrary to God's wishes not preach that slow-motion suicide by cigarette is equally unacceptable and immoral? Why would the National Council of Churches take on the issue of infant formula use in the Third World, claiming that it was life-threatening, and then turn their heads on an issue which has killed millions of Americans—and is about to kill millions of new smokers in developing nations?

The Roman Catholic publication *America* has skirted the "when-is-a-cig-a-sin" issue on a number of recent occasions, but has always left the impression that God would be vindictive only if smoking—or other things—were done to "excess."

"Consumer advocates"

The so-called consumer advocacy movement, virtually synonymous with the name Ralph Nader, has expressed outrage over what it sees as consumers being ripped off by the greedy, uncaring food, pharmaceutical, automobile, insurance and petrochemical industries. Articles emanate almost daily from the Nader-inspired Center for Science in the Public

Interest (CSPI), the Community Nutrition Institute, and other such groups, claiming that pesticides, food additives, air pollution, nuclear energy, drugs, chemical waste disposal dumps, cosmetics, coffee and candy are all killing us. Why do these great "protectors of the people" almost never speak out against tobacco?

In the 1960s, Dr. Sidney Wolfe, director of the Nader-inspired Health Research Group, was active in bringing the facts about cigarettes and health to center stage, but he no longer appears to have cigarettes on his hit list. On January 11, 1984, to observe the 20th anniversary of the first Surgeon General's report, the American Council on Science and Health sponsored a major news conference on smoking. All health groups invited to participate eagerly sent a representative, except for CSPI and the Health Research Group, both of whom even declined to submit a statement for release at the meeting.

Why the silence? Is it merely coincidental that CSPI receives substantial support from the Mary Reynolds Babcock Foundation and the Arca Foundation, family foundations of the descendants of R. J. Reynolds himself? Observers of a suspicious nature might also wonder about the issue of CSPI's *Nutrition Action* that featured a cover story on the prevention of lung cancer. While speculating about a protective action of beta-carotene (a substance found in carrots and other vegetables), it *completely ignored* the subject of cigarettes. Perhaps these so-called consumer groups are afraid that if they focused on the enormous health menace posed by cigarettes, their favorite targets such as food additives and sugar might no longer seem threatening by comparison.

On the other hand, it might be that the consumer activists are frustrated or feel that our society has gone about as far as it can in waging war against the cigarette. What may be missing, then, is momentum rather than a specific intent to avoid the topic.

Conservative leaders

Consumer liberals who harp constantly on relatively trivial or nonexistent environmental dangers are not the only ones who deserve criticism for failing to rally against cigarettes. Here are typical examples from the right side of the political spectrum:

• President Reagan (who appeared many years ago smoking in a Chesterfields ad but is a nonsmoker) has said, "Tobacco—no less than corn, wheat or soybeans—should be viewed as a valuable cash crop with an important role to play in restoring America's balance of trade."

• James Kilpatrick, writing for Universal Press Syndicate, stated that "The tobacco people are quite right. The causal relationship of cigarettes and cancer hasn't been proved, it still is only statistically inferred." On *60 Minutes* he said, "My own guess is that the anti-smoking zealots are mostly latter-day Puritans. They're like the Prohibitionists who didn't resent people's drinking half as much as they resented other people's pleasure."

• Pat Buchanan, of radio station WRC, has commented, "The American people have voted freely in the marketplace to go right on smoking—and if 50 million free people are willing to take the risks of smoking to enjoy what they see as the pleasure of smoking, who is Califano to tell them they can't."

• William Safire, *New York Times* columnist has suggested, "A little group of willful persons, representing no opinion but their own, has rendered the great smoking public helpless and contemptible. Today the smoker, tomorrow the onion eater and the day after that the person who prefers cheap perfume to taking a bath."

• William Buckley, in his book, *Jeweler's Eye,* noted "I suggest only that the government cannot, for reasons that go beyond the womb of freedom, do anything, anything at all, about smoking."

• Economist Milton Friedman wrote, "The evidence on the harmful effects of smoking, though certainly strong, is not conclusive . . . In a free society, a government has no business using the power of the law or the taxpayer's money to propagandize for some views, and to prevent the transmission of others."

• *Barron's*, a national financial weekly of distinctly conservative leanings, commented that the anti-smoking movement "began a few years ago as a seemingly well-intentioned, if disturbing effort to brainwash the citizenry into kicking the cigarette habit, thus has spiraled into a crusade as menacing and ugly as Prohibition."

• Jeffrey St. John, a Mutual Network news commentator and syndicated columnist, told tobacco executives at an industry meeting that certain "consumer groups have created a climate of intellectual terror by insisting that everything produced by the private-economic system is lethal to human life and limb."

The conservative community's near-unanimous defense of cigarettes—or at least the near-unanimous criticism of those promoting efforts to discourage smoking—is curious. While these individuals generally feel that the evidence indicating cigarettes is less than convincing, they never tell us what type of evidence they would accept.

The reluctance of conservatives to accept the need for efforts to unhook Americans from cigarettes appears to be based on three lines of thought. First, conservatives seem unwilling to accept the fact that cigarettes are so dangerous that they should be considered in a category by themselves. Second, by admitting that there is a health problem, conservatives would simultaneously be acknowledging a case of free enterprise which had gone sour. Third, conservative philosophy opposes government "do-gooding."

This philosophy maintains that government should not be responsible for protecting people from themselves, and that individuals should be free to make their own decisions without government interference. Even if one accepts these concepts, the fact is that smokers have been forcing other people to pay for the consequences of their habit. If smokers were carrying their own burden in terms of medical expenses, life and fire insurance premiums, taxes, and other costs attributable to smoking, then the arguments of the conservatives would at least be consistent. But, as noted in Chapter 15, they are not!

There is another irony in tobacco's defense by conservatives. Many who consider themselves "pro-life" when it comes to discussion of abortion do not see it as inconsistent to be in favor of—or at least not against—a product that causes widespread death.

Recommendations for action

Tobacco companies have been getting away scot-free with murder in the form of over 350,000 excess deaths each year in the United States. Tobacco industry propaganda must be challenged, and so must the industry's ability to stifle the flow of information about the hazards of smoking. If you are concerned about reducing the risks and bringing the facts to smokers, the following suggestions may be useful:

- If you (or a friend or relative) must smoke, choose a less risky way than cigarettes. Switch to a pipe or cigar and don't inhale.
- If you have friends or relatives who smoke, make sure that they know all the facts about the health dangers of smoking. Tell them, or offer them appropriate literature on the subject to read.
- Write to the FTC and your congressman, encouraging them to support efforts to promulgate information on the health risks of smoking.
- If you are a smoker, write to the president of the tobacco company whose brand you smoke. Tell him that you would feel much more comfortable smoking his brand if the ingredients were fully disclosed.
- Write to the Surgeon General's office encouraging him to continue efforts to force the tobacco industry to disclose the ingredients of cigarettes.
- Demand that your local elementary, junior high, and other high schools devote a substantial amount of time to discussing the dangers of smoking in health, and/or physical education classes.
- Write to the tobacco companies and ask what chemicals they add to their products.
- Write to the Tobacco Institute and the Chief Executive Officers of the major tobacco companies and ask them exactly what type of scientific data would convince them that their product was harmful?
- Write to your local newspaper, encouraging it to relate causes of death

to smoking habits in news articles and obituary columns, and to increase coverage of the cigarette issue.

• Keep alert to the standard rationalizations of the tobacco industry. You will find they are used again and again on television, radio and in the print media. Point out these fallacies to editors, producers and directors, and ask them to stop giving equal time for nonsense.

• Encourage feminist groups like the National Organization for Women to take an official stand on the subject of cigarettes and female health.

• Write to *Cosmopolitan, Ms., Self, Working Woman* and other feminist-oriented publications and encourage them to take on the serious subject of cigarette smoking in editorials and articles. You might recommend that all women's magazines choose a particular month to break the conspiracy of silence and write about the dangers of cigarettes. It is extremely unlikely that the tobacco companies would pull ad copy from *all* women's magazines.

• If you belong to a consumer activist group, tell its leaders that you want to hear more from them about the dangers of cigarette smoking.

• Inform conservative spokespersons that there is a solution to the cigarette problem which is compatible with the free enterprise system. Encourage them to explore routes of transferring the burden of smoking to the smoker and to stop their denials about the cigarette-disease link.

• Write to the office that coordinates the activities of any religious organization to which you belong, asking it to offer some guidance about cigarettes.

21

Smoking Cessation: An Overview

"It's easy to quit smoking. I know because I've done it thousands of times."—Mark Twain

Right now, about 56 million Americans smoke. Sixty percent of them have seriously tried to quit, and another 30 percent say they would try if there were an easy method. In all, some 90 percent of American smokers say they would like to quit. Although smoking involves a powerful chemical addiction and behavior patterns that are tough to break, some 33 million Americans—about a third of those who did smoke—*have* kicked the habit during the last 15 years.

Some smokers who have puffed for many years think that the harm is done, so there is no point in giving up the habit. This is absolutely untrue! *No matter how long a person has smoked, it is still beneficial to quit.* During the first day, nicotine and carbon monoxide levels decrease in the body, and the heart and lungs begin to repair the damage caused by cigarette smoke. "Smoker's cough" usually disappears within a few weeks, energy and endurance may increase, and taste and smell may return for foods that haven't been enjoyed for years. A decade after stopping, the risk of dying from cardiovascular disease declines to the level of the nonsmoker. The risk of lung cancer decreases, and so do the incidence of respiratory infections and the lung tissue destruction leading to emphysema. In a study of over one million ex-smokers, the death rate 10 years after quitting approached that of people who had never smoked.

The successful quitter

Success in quitting is related to personal characteristics, the length of smoking history, and the amount of daily cigarette consumption. Those smoking the longest may find it toughest, but by no means impossible, to quit.

Successful quitters make a firm personal commitment to stop. Most cite concern about health as the main reason for quitting. The Seventh-day Adventist's Five-Day Plan centers around the words, "I choose not to smoke." The American Lung Association's self-help manual urges participants to sign a special stop-smoking contract. Reputable hypnotists will not accept clients who lack a strong personal desire to stop. Attempts to quit only to please other people are usually doomed to failure. Tom Wicker, syndicated columnist for *The New York Times* and a former smoker, summed this up well when he wrote: "If you *want* to stop smoking, you can; if you merely think you *ought* to, you're kidding yourself."

No smoking cessation method fits everyone. Surveys show that 90 to 95 percent of smokers prefer to stop on their own or by using printed instructions, guides or videos. Others need informal group support or counseling. Most who succeed are committed to personal change and are well aware of their reasons for wanting to stop. Studies show that these people are open to trying any of a variety of cessation programs rather than any particular method. It is essential for smokers to choose the method that conforms most closely to their personal needs. Timing is also important. Trying to break the habit in the middle of an important business or family crisis won't enhance one's chances for success.

Withdrawal symptoms

Simply put, withdrawal is the process of getting used to doing without something the person has depended on for a long time. Withdrawal symptoms indicate that the body is taking the first big step towards breaking dependence upon cigarettes. When a smoker quits, the body no longer has the stimulant nicotine to rely on. Certain habits also must change.

Symptoms vary from person to person, both in character and in duration. One researcher has said, "The impression might be drawn that every smoker, upon withdrawal from tobacco, becomes irritable and anxious and is unable to think, work, sleep, drive, or carry on normal social discourse for want of a cigarette. Fortunately, this is not the case." According to the American Cancer Society, most withdrawal symptoms disappear within a week or two.

Common symptoms during the withdrawal period include:

- *Craving* for cigarettes, one of the most common symptoms, is experienced by 90 percent of smokers who stop. Craving can become noticeable within two hours after the last cigarette and usually reaches its peak within 24 hours. It then gradually declines over a period of days or weeks. About half of quitters who relapse cite craving as a major factor.
- *Nervousness and restlessness* are related to withdrawal from nicotine. Drinking lots of water or fruit juices helps to flush nicotine out of the body. Staying off caffeine may help because it can cause nervousness.

However, some swear that drinking lots of coffee actually helped them to quit.

- *Cough.* Smoking paralyzes the hairlike structures (cilia) that normally clean the lungs by "sweeping out" foreign matter. (This is the reason why smokers have more respiratory infections than nonsmokers.) Increased coughing may occur as the cilia regain full function. Coughing eventually subsides as natural lung function takes over.
- *Headache.* According to the American Lung Association, extra rest, deep breathing exercises, or increased exercise may help in dealing with this symptom.
- *Tingling or numbness* of the arms and legs may occur temporarily as circulation to these areas improves.
- *Tiredness.* Smokers may be used to a higher level of arousal due to the effects of nicotine. Cutting out nicotine is a let-down to the body. Extra exercise and an extra hour of sleep are recommended.
- *Lack of concentration.*
- *Slight sore throat.* Tobacco smoke irritates the throat but also numbs it. Discomfort is felt as the numbness wears off and the throat begins to heal.
- *Constipation* occurs in some quitters. Increased dietary fiber can help. Otherwise a physician should be consulted.

What about weight gain?

- *Weight gain* can occur, but it is avoidable. Sixty percent of female smokers and 47 percent of male smokers say they are reluctant to quit for fear of weight gain. Actually, about a third gain weight, a third stay about the same, and a third lose weight while quitting—but most of the gainers don't gain very much. There is no significant health risk associated with a small weight gain. Only a gain of more than 75 pounds would offset the benefits of quitting for a normal smoker.

Quitters often try to nurse the symptoms of withdrawal by snacking more than usual or rewarding themselves for cigarettes not smoked by consuming extra-rich treats. Some experience an improved sense of taste which increases their desire to eat. Others feel a need to put something in their mouths to replace cigarettes.

Since the majority of quitters gain very small amounts or actually lose weight, a few intelligent controls can be quite effective; weight gain during smoking cessation can be avoided. Almost every smoking cessation program includes ideas and suggestions to help quitters avoid weight gain. Most stress stepped-up exercise, low-calorie snacks, and a balanced diet consisting of three nutritious meals per day. A smoking cessation program that does not deal seriously with the possibility of weight gain would be overlooking an important withdrawal factor.

Smoking Cessation: An Overview   215

Many people wonder whether it is better to quit suddenly ("cold turkey") or make a gradual reduction. Although millions have had success with either approach, studies indicate that stopping suddenly gets the withdrawal process over with more quickly. Cutting down somewhat before stopping completely may reduce the intensity of withdrawal symptoms. However, too much tapering off will cause withdrawal symptoms to occur between cigarettes in addition to those that must be faced after smoking is stopped.

Self-help publications

Smokers quitting on their own can benefit from a variety of manuals designed especially for people who prefer to go it alone. These publications encourage the smoker to plan ahead, to anticipate the problems of withdrawal, timing, anxiety and weight gain.

The American Lung Association's *Freedom from Smoking in 20 Days* offers a down-to-earth approach designed to increase the smoker's confidence. Organized into daily segments, the booklet is attractively illustrated and describes each step of the way. Days 1 to 7 are dubbed "Preparation," during which the smokers explore their reasons for smoking, their smoking habits and the way to set up a system of rewards for making progress in the program. A 9-day period follows in which the smokers are asked to change their pattern of smoking. Quitting can be done "cold turkey" or by eliminating a few cigarettes at a time. During this phase, the individual learns how to deal with the symptoms of withdrawal. The final 4 days are designed to reinforce the nonsmoking habit by having the nonsmoker think up self-rewards, exercise more often, and recall the reasons for wanting to quit.

A follow-up booklet, *A Lifetime of Freedom from Smoking*, reminds the ex-smoker of the things that trigger the urge to smoke and how to best cope with them. It also covers how to deal with tension, social situations, and weight gain. The ALA booklets can be obtained from a local ALA chapter for $10 or less.

The American Cancer Society's "I Quit Kit" ($1.50) utilizes many of the same techniques as the ALA materials. It contain educational materials and some novelties such as a record with songs and humor about smoking and instructions about breathing exercises. Other sections contain a poster and buttons exclaiming "I'm a Quitter" and "I Quit." According to the ACS, the kit is a "hot item" and is being ordered in quantity by its local divisions. The Society also publishes a free-of-charge pamphlet version of the kit called the "Quitter's Guide."

The National Cancer Institute has been providing physicians and dentists with free "quit kits" for distribution to their patients. The kits contain instructions to the doctor plus supplies of two leaflets and a booklet. One leaflet, *What Happens after You Quit,* summarizes the benefits of quitting. The other leaflet, *Why Do You Smoke?,* contains a self-test designed to help smokers choose the best method of quitting for themselves. The booklet *Calling It Quits* provides a plan for quitting and has many practical tips.

Self-help guides are important because they enable quitters to maintain privacy and independence while providing a framework for anticipating problems and planning ahead.

Over-the-counter products

Several nonprescription products are available at local drugstores. Some are designed to reduce the amount of tar and nicotine in cigarette smoke while others are intended to deter smoking.

Filters are being marketed which, when attached to the unlit end of a cigarette, substantially reduce the amount of tar and nicotine inhaled by the smoker. For example, Water Pik's "One Step at a Time" includes a set of four reusable filters designed to reduce the tar and nicotine content of smoke by 25, 50, 60 or 90 percent. The idea is for the smoker to progress to a stronger and stronger filter and then to kick the habit after becoming accustomed to the most effective filter.

Unfortunately, studies have shown that even though such filters are effective in reducing tar and nicotine in cigarette smoke, they have not been effective in helping smokers to stop. One study published in *American Pharmacy* found that no smokers had quit after an 8-week period. In fact, many people smoked more to maintain their nicotine level—the same thing that often happens when people switch from high- to low-tar/nicotine cigarettes. Another study found that many subjects who progressed to the fourth and strongest filter were unable to take the final step to quit.

Smoking deterrents are drug products that either alter the tobacco taste so that smoking becomes less pleasant or substitute a nicotine-mimicking drug. Deterrents are supposed to produce an effect that alters the smoker's habit or addiction.

The FDA is currently reviewing the safety and effectiveness of smoking deterrents. A special panel of experts has judged that 43 of the 45 active ingredients in these products are not generally recognized as safe or effective. The other two ingredients—lobeline and silver acetate—are still under study because the panel found insufficient evidence on which to base a decision.

Studies on lobeline sulphate (contained in Bantron, Nikoban, Lobidan, Tabusine, and lobeline hydrochloride) have shown conflicting results. In one study of 200 chronic smokers, more than 80 percent of the individuals testing the drug had stopped smoking at the end of 5 to 6 days. Less than 10 percent using a placebo (inactive dummy pill) quit. But only 2 out of 10 other controlled studies found lobeline products more effective than a placebo in helping the subjects to stop smoking.

Silver acetate is used as an ingredient in astringent mouthwashes. The silver salts are said to affect the mucous membranes of the mouth and throat so that an unpleasant metallic taste occurs in response to smoking. These uncomfortable effects can last up to 4 hours, thereby making smoking less desirable and reducing the number of cigarettes smoked per day. So far three controlled studies have not demonstrated any effectiveness for silver acetate. The FDA is conducting further studies.

Action by the FDA on the status of all over-the-counter smoking deterrents is expected soon.

Nicotine chewing gum

Now for some encouraging news. Nicotine administered by a route other than smoking may be effective in helping smokers kick the habit for good. In Sweden, 17 years ago, a nicotine chewing gum was developed as an aid to smoking cessation. It is also available commercially in Great Britain, Canada, Austria and Germany. Early in 1984, the gum was approved for use in the United States under a physician's prescription. It is currently marketed under the brand name Nicorette by Merrell Dow Pharmaceuticals Inc., a subsidiary of the Dow Chemical Company.

An average cigarette contains a little more than 1 milligram (mg) of nicotine. Nicotine chewing gum is produced from a natural extract of the tobacco plant, with each piece containing 2 mg of nicotine. The nicotine is slowly absorbed through the lining of the mouth over a 20 to 30-minute period. Studies have shown that chewing one piece of 2 mg gum per hour produces blood nicotine levels comparable to those obtained with hourly cigarette smoking. Any nicotine that is swallowed has little effect on the body.

The gum's purpose is twofold—to provide a substitute for oral activity and to prevent nicotine withdrawal symptoms. This allows smokers to tackle quitting in two stages. First, the psychological or behavioral habits can be broken without the smoker having to suffer the discomforts of nicotine withdrawal. Second, after a few months, the use of nicotine gum is gradually tapered off.

Recent studies using nicotine gum together with behavioral therapy have been relatively favorable. One British study demonstrated that 38 percent of 69 hard-core smokers given the gum were abstinent after one year, compared to only 14 percent of 49 who had received only psychological treatment. In another study by the same researchers, 47 percent of subjects who had received the gum were abstinent after one year compared to only 14 percent who received placebo gum containing nicotine in a nonabsorbable form. The latter study also involved counseling—one 1-hour session per week for six weeks.

An FDA Drug Abuse Advisory Committee which reviewed the evidence on nicotine gum concluded that it does increase the likelihood of smoking cessation *when used in conjunction with an acceptable counseling program.* Merrell Dow has prepared a complete package of information directed at physicians and patients which outlines counseling techniques and product usage. Doctors are supplied with an 8-step guide describing in detail how to select and prepare patients for using the gum. Patients receive a 24-page booklet similar to the self-help manuals discussed earlier in this chapter. There is also in the package a business reply card to send to Merrell Dow, which then sends the patient two newsletters about smoking cessation and nicotine gum treatment. Doctors are urged to provide at least one follow-up visit per month over the treatment period (usually 3 to 4 months). Those who cannot provide educational support are urged to refer their patients to an agency that can. A 96-piece package of Nicorette costs about $20.

The gum is expected to be well received, but experts worry that physicans may not prescribe it properly or provide the necessary counseling. It seems likely that the ultimate utility of nicotine chewing gum will be determined by the way in which it is used; but for many smokers, there may be light at the end of the tunnel.

Medical supervision

Physicians are in a unique position to encourage their patients who smoke to stop. A medical opinion carries a great deal of weight. The National Cancer Institute has developed a pamphlet, "Helping Smokers to Quit—A Guide for Physicians," for doctors who are unsure about the type of advice to give their patients.

What can a smoker expect the doctor to do? The answer is not simple. A doctor can't merely write a prescription for a miracle cure for the smoking habit. Here are some guidelines suggested by the National Cancer Institute. A doctor should:

- Attempt to learn everything possible about the patient's motivation for quitting and help the patient to verbalize this.
- Help smokers determine why they smoke. Often the answer to this question will increase the chances of success.
- Introduce the various approaches to quitting but urge the smoker to choose the right method.
- Provide encouragement, understanding, and follow-up during the first several weeks after quitting.

Hypnosis

More than a quarter of a century has elapsed since the American Medical Association accepted hypnosis as a legitimate tool in medical treatment. During those years hypnosis has become popularized as a technique used to help smokers quit. Hypnosis is not a panacea, but it may be worth a try for quitters who have tried other methods and have failed several times.

Although hypnotists tend to develop their own techniques, many concentrate on focusing the smoker's attention on smoking triggers—those everyday occurrences that cause a smoker to want to light up. In this way, smoking can be treated behaviorally as a variety of habits.

Although some reports on the use of hypnosis for smoking cessation claim a high rate of success (claims range from 15 to 80 percent), it is hard to be sure how to interpret their results. Since people willing to pay $200 to $400 for a series of hypnotic sessions are likely to have above-average motivation to quit, it would not be surprising if many of them are successful. Moreover, paying a large sum may help induce people to try harder to quit so they can feel that they got what they paid for.

Choosing a hypnotist can be tricky because some of them are quacks. Names of reputable practitioners can be obtained from one's family physician or by contacting any of the following:

American Society of Clinical Hypnosis
2250 East Devon, Suite 336
Des Plaines, IL 60018

Society for Clinical and Experimental Hypnosis
129A King Park Drive
Liverpool, NY 13088

International Society for Professional Hypnosis
218 Monroe Street
Boonton, NJ 07005

The International Society for Professional Hypnosis warns against individuals who claim to have degrees in hypnosis or "hypnology," or who claim to have licenses as hypnotists. Credentials of this sort are bogus. The Society also cautions against hypnotists who advertise 1-session cures.

Acupuncture

For cigarette smokers, acupuncture treatment involves the insertion of needles into various parts of the ear. This is usually done at the rate of two sessions per week for two or three weeks. Between treatments the needles are left in the ear and the quitter stimulates these when the urge to smoke is felt. This supposedly makes cigarettes taste funny and changes the urges of the would-be quitter. Most practitioners will refuse to continue treatment if acupuncture is unsuccessful after seven visits. Acupuncturists usually charge at least $35 per visit, placing the average treatment in the $200 range.

Acupuncture for smoking cessation has a low success rate—only about

15 percent. Acupuncturists themselves often claim rates of 80 percent or higher. Figures this high are unsubstantiated by scientific research and should be regarded with great suspicion.

Group methods

Scientifically-planned cessation programs are available in many communities from such nonprofit organizations as the American Cancer Society, American Lung Association, Seventh-day Adventist Church and American Health Foundation. Other organizations such as SmokEnders, Smoke Watchers, and the Schick Clinics, operate for profit. Local hospitals may also have smoking cessation clinics.

Group methods add additional forms of support that self-help methods lack. First, experienced and specially trained group leaders conduct these programs in a highly professional manner. Second, the participants in group programs can share problems, successes and encouragement. Here is a brief analysis of the leading methods.

• *The Five-Day Plan* was devised in the early 1960s by two physicians who believed that simple techniques could be used to overcome the physiological, psychological and behavioral aspects of cigarette addiction. The plan is supported by the Seventh-day Adventist Church and is available to the public free of charge or for a nominal fee. The plan has demonstrated versatility and has been offered on commuter trains and on television, in hospitals, factories, prisons, and government offices, and by physicians, health care agencies, civic organizations, and the military service. It is also available through films and cassettes. Studies of the Five-Day plan have shown abstinence rates between 16 and 40 percent after one year.

In the program, quitters are asked to drink lots of water and fruit juices. According to Dr. Ihor Bekersky, a drug researcher for a major pharmaceutical company who also conducts Five-Day programs, "Drinking water actually flushes the nicotine out of the body . . . Nicotine is reabsorbed into the blood stream from the kidneys and urinary bladder. Increasing consumption of liquids counteracts this."

Why do some smokers still experience cravings for nicotine several weeks after quitting? Says Dr. Bekersky, "We believe that nicotine somehow binds with or alters the special neurochemical receptors that are important to the body's response to catecholamines." Catecholamines are a family of chemicals produced by the body to enable it to respond to stress and to help send nerve impulses. It may take several weeks for these receptors to return to normal. Until then, explains Bekersky, withdrawal symptoms may be felt.

The Five-Day Plan also concentrates on associated factors. For example, smokers often associate a cup of coffee with a cigarette. According to Vincent Gardner, M.D., a physician who has conducted 25 programs a year for the past ten years, the plan allows no coffee or cola beverages, since caffeine puts a strain similar to nicotine on the catecholamine system. (Others question this point.)

"In the Five-Day Plan you are in control," says Dr. Gardner. "The nonsmoking image is constantly being reinforced by other health habits." For example, smokers often light up after a meal. The Five-Day Plan prescribes going outside and taking a walk instead of smoking. Other suggested techniques are warm baths or showers for relaxation (it's also quite difficult to smoke in the shower), adequate rest, use of a "buddy system," and eating less spicy or greasy foods which often stimulate craving. Vitamin supplements are recommended, especially B_1 (thiamine) which supposedly calms the nerves. There is no scientific evidence to support this recommendation. The best nutrition advice is simply to eat three moderate-sized well-balanced meals daily and to limit snacking.

• *FreshStart* is offered by the American Cancer Society (ACS). The newly developed program replaces the older "Helping Smokers Quit Clinics" and is conducted by extensively trained volunteers. The clinic meets twice weekly for two weeks. The Society claims a success rate of 25 to 30 percent after one year.

The program's first session is devoted to understanding "Why and how I smoke and how smoking affects me." Smokers take a test designed to show if they are chemically addicted, habituated to smoking, or psychologically dependent. "It is not necessary to get rid of the desire to smoke totally before you stop," states the manual for the clinic. "Often the desire will only go away after one has stopped smoking." Session 2 introduces stress-management techniques such as deep breathing exercises and assertiveness training. Session 3 covers obstacles such as weight gain, and the final session talks about how to avoid returning to smoking.

Preparing to quit is de-emphasized, while planning ahead to stay off cigarettes is emphasized. The program allows a choice between gradually decreasing cigarette consumption or quitting "cold turkey." The sessions also focus on what the ACS considers the three most important aspects of smoking: chemical addiction, habit, and psychological dependency.

Information about FreshStart can be obtained from any local ACS chapter. There is a nominal fee to cover expenses.

• *Freedom From Smoking clinics*, offered by the American Lung Association (ALA), follow the same themes presented in its self-help manuals. Test results were used to design what it considers to be the most effective format: seven sessions conducted over a seven-week period. The first three sessions make smokers aware of the events that trigger smoking and how to cope with them. Participants are encouraged to develop personal plans of action to deal with the urge to smoke. As in other methods, a "Why Do You Smoke?" test is given and the buddy system is introduced. To help maintain the nonsmoking lifestyle after the quit night (the third session), the program covers such topics as the benefits of quitting, relaxation skills, exercise and weight control methods, and ways to deal with social situations that may cause relapse.

Not all local chapters of the ALA offer the Freedom From Smoking clinics as yet. However, specially trained ALA personnel offer the program nationwide. A fee of about $35 is currently required. Based on its own studies, the ALA claims a quit rate of 22 percent immediately following the clinics, which rises to about 40 percent after one year. This suggests that some smokers who failed to quit during the 7-week clinics implemented their information months afterwards.

- *SmokEnders* is a commercial agency which got its start in New Jersey in 1969. The technique is highly structured, systematic, and does not depend on the use of scare tactics. SmokEnders classes can be found across the country in most major cities. Tuition costs several hundred dollars, but there is a 25% discount if a smoker taking the course enrolls a partner as well. The program is composed of ten weekly sessions, the first one of which is free. During the first four weeks participants learn techniques to help them postpone smoking a cigarette so that by the fifth week they can stop smoking entirely. The last four weeks are devoted to encouragement, support and discussions of the problems that participants encounter. Independent studies of the SmokEnders program show a success rate of about 27 percent.

- *Schick Clinics*, available on the West Coast, utilize a technique called adverse conditioning through the use of rapid-puffing techniques and mild electric shocks; the program literally makes the smoker sick of smoking. Educational and counseling services are also provided in the 5-day clinic which can cost up to $750. Schick claims a 53 percent success rate after one year, but has refused to open itself to independent confirmation.

- *Smoke Watchers* was the first commercial program in the United States. They claim a 37.5 percent success rate using techniques of gradual withdrawal and education. The cost is an initial membership fee of about $25 plus a small additional fee per session.

Success rates

Success rates should be interpreted with caution. While some groups have opened themselves up to scientific investigation, others have not. To be valid, a comparison of rates must take into account how they were determined—how "success" was defined, when the measurements were taken, how the data were obtained, and so on. A number of factors can bias success rates to the high side. For example, questionnaires sent by mail are more likely to evoke responses from those who are successful than from those who are not.

On the other hand, the fact that most reputable programs have success rates in the range of 20 to 40 percent should not be interpreted to mean that most smokers who attempt to quit are likely to fail. Dr. Stanley Schachter, a psychology professor at Columbia University, studied two random population groups (which included many people who had tried on their own) and found a higher percentage of those who tried to stop *were* successful.

Advice to businesses

Americans spend hundreds of billions of dollars yearly on health care. Insurance industry estimates show that employers pay about half of these costs. The American Lung Association reports:

1. Direct health care costs for smoking-related illnesses . . . total an estimated $13 billion every year.
2. Lost productivity and wages due to these illnesses account for an additional $25 billion yearly loss.
3. Each year, more than 80 million work days are lost to smoking.
4. Smokers average 35 to 45 percent more absenteeism than nonsmokers.
5. Costs to employers of an individual employee who smokes have been placed in the range of $400-$600 per year.

The ALA and the *American Health Foundation* (AHF) provide free consultation to companies that wish to establish smoking cessation programs for their employees. Company representatives can travel to AHF offices in New York City for a full-day training session on needs assessment and strategy planning for smoking cessation. Company representatives may also become qualified to teach the AHF's own smoking cessation program by participation in a 3-day training session. The ALA conducts similar services and also offers clinics in the workplace and training of personnel to conduct clinics.

Summing it up

Whether a smoker decides to go it alone or to make use of one of the many aids and programs available, one thing is certain: Kicking the habit is not a breeze, but there are some very good ways of making it easier.

To the smoker the course of action should be clear. Make the decision to quit. Make the commitment to quit. Develop a plan or find a program that's right. Carry out the plan and stick to your guns! Addresses of helpful agencies are listed in Appendix B of this book.

A final comment

Dr. Frank A. Oski is Professor and Chairman of the Department of Pediatrics at University Hospital in New York. Recently, after surviving his first heart attack at the age of 51, he shared these thoughts with readers of *The New York Times*:

> Will I miss the late night trips to find a store still open and selling cigarettes? Will I miss rummaging through ashtrays to find the longest butt that is still smokable?
>
> Only time will tell. Not smoking may give me the time to find out.
>
> Was it easy to stop? Sure. Here is all you have to do. First, experience a severe crushing pain under the breastbone as you finish a cigarette. Next, have yourself admitted to a coronary care-unit and stripped of all your clothing and other belongings. Finally, remain in the unit at absolute bed rest for four days while smoking is prohibited. This broke my habit. See if it works for you.

Amen.

Appendix A: Tobacco's Industrial Network

MAJOR U.S. CIGARETTE MANUFACTURERS

American Brands, Inc.
245 Park Avenue
New York, NY 10167

Brown & Williamson Tobacco Corp.
P. O. Box 35090
Louisville, KY 40232

Liggett & Myers, Inc.
West Main & Fuller Streets
P. O. Box 1572
Durham, NC 27702

Lorillard
Loews Theatres, Inc.
666 Fifth Avenue
New York, NY 10103

Philip Morris, Inc.
100 Park Avenue
New York, NY 10017

R. J. Reynolds Tobacco Co.
401 N. Main Street
Winston-Salem, NC 27102

TOBACCO'S INDUSTRIAL NETWORK

American Brands
American Cigar: cigars and smoking tobacco
American Tobacco Company: cigarettes and smoking tobacco
Acme Visible Records, Inc.: record retrieval and storage systems, computer retailing
Andrew Jergens: soap, lotions, shampoo
Acushnet Company: golf equipment, rubber parts for cars and oil rigs

Duffy-Mott Co.: juices, molasses
Franklin Life Insurance Company: Life insurance—(offers discount to nonsmokers!)
Gallaher Limited: cigarettes, cigars, pipe tobacco, optical retailing pumps and valves, retailing, office products distribution
James B. Beam Distilling Co.: liquor, mixers, trophy bottles and ceramic products
Master Lock Company: padlocks
MCM Products, Inc.: knives, shears, scissors, light bulbs, auto body and riveting products
Sunshine Biscuits, Inc.: biscuits, crackers, snacks
Swingline, Inc.: stapling and fastening equipment, pneumatic tools
Wilson-Jones Company: binders, cabinets, office forms, etc.; information retention and retrieval systems, corrugated storage systems

BATUS (BAT in U.S.)
Brown & Williamson Tobacco Corporation: cigarettes (domestic & international)
Export Leaf Tobacco Company: purchases, processes and stores U.S. tobaccos, supplies to subsidiaries and affiliates
Appleton Papers: pulpmaking, papermaking and converting
Gimbels: department stores
Kohl's Department Stores: department stores
Kohl's Food Stores: grocery stores
Marshall Field and Company: department stores
Saks Fifth Avenue: department stores
Thimbles: specialty fashion stores

Grand Metropolitan
Liggett & Myers, Inc.: cigarettes, tobacco products, spirits and wines, pet foods, soft drinks, sporting goods, food ingredients
Express Dairy: manufactures dairy products and related caterer's food lines, liquid milk and other fresh foods, production of tomatoes
Grandmet Hotels: hotels, catering services, steakhouses, hospital management
Intercontinental: hotels
International Distillers & Vintners: liquor marketing and production
Mecca Leisure: bingo social clubs, entertainment nightspots, ice skating rinks, "high class" restaurants, betting offices, casinos, catering services
Watney Mann & Truman Brewers: beer, pubs, soft drinks

Loews Corporation
Lorillard: cigarettes, chewing tobacco
Bulova: watches, clocks, military defense
CNA Financial: insurance underwriting, other financial services
Loews Hotels: hotels in Bahamas, France, etc.
Loews Theatres: movie theatres

Philip Morris Incorporated
Philip Morris U.S.A.: cigarettes
Philip Morris International: cigarettes
Philip Morris Industrial: pulp-based and chemical products
Miller Brewing Company: beer
Mission Viejo Company: housing developments
Seven-Up Company: 7UP

R.J. Reynolds Industries, Inc.
R.J. Reynolds Tobacco Company: cigarettes, smoking tobacco, plug chewing tobacco, pouch chewing tobacco, little cigars
R.J. Reynolds Tobacco International: international tobacco
Aminoil USA, Inc.: domestic petroleum exploration and production, natural gas processing, geothermal steam development and supply, foreign petroleum
Del Monte Corporation: canned foods, frozen foods, Hawaiian foods, Chinese foods, Mexican foods, fresh fruits
R.J. Reynolds Development Corporation: foil and aluminum products, protective film wrap, manages food service facilities for businesses and institutions and provides contract maintenance, engineering and security systems
Sea-Land Industries Investments, Inc.: containerized ocean and overland transportation

Appendix B: Helpful Organizations

Action on Smoking and Health (ASH), 2013 H St., N.W., Washington, DC 20006. A nonsmokers' rights organization which publishes a bimonthly newsletter and interim news bulletins, files lawsuits, and sells a variety of no-smoking campaign materials.

Air Travelers Safety Association, 2908 Patricia Drive, Des Moines, IA 50322. Involved in trying to prohibit smoking on airplanes.

American Cancer Society, 777 Third Ave., New York, NY 10017. Publishes educational materials and operates local smoking cessation programs.

American Council on Science & Health, 1995 Broadway, New York, NY 10023. Publishes pamphlets and a bimonthly newsletter which contains articles about the problems of tobacco products.

American Dental Association, 211 E. Chicago Ave., Chicago, IL 10019. Publishes educational materials about oral problems caused by tobacco products.

American Health Foundation, 320 E. 43rd St., New York, NY 10017. Publishes educational materials and consultation to companies wishing to establish smoking cessation programs.

American Heart Association, 7320 Greenville Ave., Dallas, TX 75231. Publishes educational materials.

American Lung Association, 1740 Broadway, New York, NY 10019. Publishes educational materials, operates local smoking cessation programs, and offers training to companies that wish to implement smoking cessation programs.

American Medical Association, 535 N. Dearborn St., Chicago, IL 60610. Publishes educational materials.

Coalition on Smoking or Health, 419 7th St., N.W., Suite 401, Washington, DC 20004. Publishes educational materials.

Californians for Nonsmokers' Rights, 2054 University Ave., Berkeley, CA 94704. Educational and political action; publishes a quarterly newsletter.

Appendix B 229

Doctors Ought to Care (DOC), 302 Turner McCall Blvd., Rome, GA 30161. Engages in public education and other activities to focus public attention on the problems of smoking.

Group to Alleviate Smoking Pollution (GASP), P.O. Box 682, College Park, MD 20740. Publishes educational materials and a newsletter.

National Interagency Council on Smoking and Health, Center for Health and Safety Studies, Office of Publications and Editorial Services, Dept. of Health and Safety Education, HPER Bldg. Room 116, Indiana University, Bloomington, IN 47405. Publishes the *Smoking and Health Reporter,* a quarterly newsletter.

Office on Smoking and Health, U.S. Dept. of Health & Human Services, Room 1-58, 5600 Fishers Lane, Rockville, MD 20857. Publishes materials on the problems of smoking and on smoking cessation.

Schick Laboratories, 1901 Avenue of the Stars, Suite 1530, Los Angeles, CA 90067. Operates local smoking cessation programs.

Seventh-day Adventist Church, Narcotics Education Division, 6840 Eastern Ave., N.W., Washington, DC 20012. Operates local smoking cessation programs.

SmokEnders, 3708 Mt. Diablo Blvd., Lafayette, CA 94549. Operates local smoking cessation programs.

World Health Organization, Regional Office for the Americas, 523 W. 23rd St., N.W., Washington, DC 20037

Many of these organizations have branches listed in local telephone directories.

Appendix C: Recommended Reading

A Smoking Gun is based on a review of thousands of articles and books. Listed below are some key references for those who wish to explore the literature on tobacco and health more deeply.

BOOKS

American Cancer Society: *The Dangers of Smoking, the Benefits of Quitting,* New York, 1982.
Califano, Joseph A. Jr.: *Governing America,* Simon and Schuster, New York, 1981.
Diehl, Harold S.: *Tobacco & Your Health: The Smoking Controversy,* McGraw-Hill, New York, 1969.
Dillow, Gordon L.: *The Hundred-Year War Against The Cigarette,* American Heritage Publishing Co., Inc., 1981.
Fritschler, A. Lee: *Smoking and Politics,* Prentice-Hall, Inc. Englewood Cliffs, N.J., 1969.
Jacobson, Bobbie: *The Lady Killers,* Pluto Press Limited, London, 1981.
Lewine, Harris: *Goodbye to All That,* Chaucer Press, 1970.
National Institute of Drug Abuse: *Research on Smoking Behavior,* U.S. Government Printing Office, Washington, D.C., 1977.
Neuberger, Maurine: *Smoke Screen,* Prentice-Hall, Inc., Englewood Cliffs, N.J., 1963.
Skinner, Willian Iverson: *Tobacco and Health (The Other Side of the Coin),* Vantage Press, New York, 1970.
Sobel, Robert: *They Satisfy,* Doubleday, Garden City, New York, 1978.
Troyer, Ronald J. and Markle, Gerald E.: *Cigarettes: The Battle Over Smoking,* Rutgers University Press, New Brunswick, NJ, 1983.
Wagner, Susan, *Cigarette Country,* Praeger Publishers, New York, 1971.
Whiteside, Thomas, *Selling Death,* Liveright, New York, 1971.
Winter, Ruth: *The Scientific Case Against Smoking,* Crown Publishers, Inc., New York, 1980.

OTHER PUBLICATIONS

American Council on Science and Health: *Cancer in the United States: Is there an Epidemic?,* 1983.
Arkin, Aaron and Wagner, David H.: Primary carcinoma of the lung. *Journal of the American Medical Association,* 106:587-591, 1936.
Babyak, Blythe. Joe Califano's anti-cigarette campaign: all smoke and no fire. *Washington Monthly,* July/August, 1978, pp. 32-36.

Bennett, William: The nicotine fix. *Harvard Magazine,* July-Aug. 1980, pp. 10-14 and Sept./Oct. 1980, pp. 12-15.
Blair, Gwenda: Why Dick can't stop smoking, *Mother Jones* IV(1):30-32, 1979.
Bock, Fred G.: Nonsmokers and cigarette smoke: a modified perception of Risk, *Science* 215:197, 1982.
Business and Society Review: Why we accept cigarette ads, Winter 1977-1978. p. 60.
Cameron Charles: Lung cancer and smoking, *Atlantic* 197(1):71-75, 1956.
Campbell, Alexander M.: Excessive cigarette smoking in women and its effect upon their reproductive efficiency, *Journal of the Michigan State Medical Society* 34:146, 1935.
Cancer of the Lung—Case No. 248301. *Newsweek,* June 22, 1959, pp. 51-55.
Catch 'em on the campus: *The Nation* 191(15):339, 1960.
Christen, Arden G. and Glover, Elbert: Smokeless tobacco: seduction of youth, *World Smoking and Health,* Spring 1981, pp. 20-23.
Cigarettes and lung cancer, *British Medical Journal,* May 19, 1956, p. 1157.
Cigarettes, cancer and morals: *America* 90(3), 1953.
Uncle Sam: tobacco salesman, *Consumer Reports,* March 1966.
Dale, Kristan C.: ACSH survey: which magazines report the hazards of smoking? *ACSH News & Views* 3(3):1, 8, 1982.
Diehl, Harold S.: Smoking out the villain, *New York Times Book Review,* November 24, 1963.
Doll, Richard and Hill, A. Bradford: The morality of doctors in relation to their smoking habits, *British Medical Journal* 1:1451-1455, June 26, 1954.
Doll, Richard and Hill, A. Bradford: Smoking and carcinoma of the lung. *British Medical Journal,* September 30, 1950, pp. 739-748.
Doll, Richard and Hill, A. Bradford: A study of the aetiology of carcinoma of the lung, *British Medical Journal,* December 13, 1952, pp. 1271-1286.
Drew, Elizabeth B.: The cigarette companies would rather fight than switch, *New York Times Magazine,* May 4, 1969, pp. 36-37, 129-133.
Dungal, Neils: Lung carcinoma in Iceland, *Lancet,* August 12, 1950.
Eckholm, Erik: The natural history of tobacco, *Natural History* 86(4):22, April 1977.
English, John P. and Berkson, Joseph: Tobacco and Coronary Disease, *Journal of the American Medical Association* 115:1327-1329, 1940.
Field, Sydney S.: Cigarettes—and sudden death, *Reader's Digest,* May 1976.
Field, Sydney: What smoking does to women, *Reader's Digest,* Jan 1976.
Foote, Emerson: Advertising and tobacco, *Journal of the American Medical Association* 245(16):1667-1668, 1981.
Garfinkel, Lawrence: The impact of low tar/nicotine cigarettes, *World Smoking and Health* 5(2):4-8.
Garrett, Nancy: Smoking now and then, *The Canadian Nurse,* Nov. 1973, pp. 22-26.
Graham, Evarts Ambrose: Primary cancer of the lung with special consideration of its etiology, *Bulletin New York Academy of Medicine,* 27(5):261-276, 1951.
Hochschild, Adam: Shoot-out in Marlboro Country, *Mother Jones,* Jan. 1979, pp. 33-34.
Hoffman, Frederick L.: Cancer of the lungs, *American Review of Tuberculosis* 19(4):392-406, 1929.
Hospital white, *The Nation* 186(3):43, 1958.
Huebner, Albert: Making the Third World Marlboro Country, *The Nation,* June 16, 1979, pp. 717-719.

Hymowitz, Norman: Personalizing the risk of cigarette smoking, *Journal of the Medical Society of New Jersey* 77:579-582, 1980.
If tobacco were spinach, *The Nation* 197(18):359, 1962.
Kupferberg, Seth: How to make the tobacco companies pay for cancer, *Washington Monthly* 9(4):56-60, 1977.
Little, Clarence Cook: The public and smoking: fear or calm deliberation, *The Atlantic,* Dec. 1957, pp. 74-76.
Lilienfeld, Abraham M.: The case against the cigarette, *The Nation* 194(13):277-280, 1962.
Mooney, Hugh J.: What the cigarette commercials don't show, *Reader's Digest,* Jan. 1968, pp. 71-74.
Miller, Lois Mattox and Monahan, James: The cigarette industry changes its mind, *Reader's Digest,* July 1958, pp. 35-41.
Miller, Lois Mattox and Monahan, James: To the cigarette makers: just the facts please, *Reader's Digest,* Nov. 1966, pp. 61-67.
Miller, Lois Mattox and Monahan, James: Cigarettes—tried and found guilty, *Reader's Digest,* April 1964, pp. 71-76.
Mosher, Beverly and Sheridan, Margaret J.: Conspiracy of silence? *ACSH News & Views* 1(2):1, 14, 15, 1980.
Muller, Mike: Tobacco and the Third World: tomorrow's epidemic? *War on Want,* London, 1978.
The People, their health and their Congress, *The New England Journal of Medicine* 28:214-215, July, 1980.
Cyclamates and cigarettes, *The New Republic* 161(18):9-10, 1969.
New York State Journal of Medicine, 83(13), 1983.
No tar test on Barclay, *New York Times,* June 25, 1982.
A frank statement to cigarette smokers, *New York Times,* Jan. 4, 1954.
Neuberger, Richard L.: Cigarettes and price supports, *Christian Century LXXIV* (36), 1957.
Norr, Roy: Cancer by the carton, *Reader's Digest,* Dec. 1952: pp. 7-8.
Ochsner, Alton: The health menace of tobacco, *American Scientist* Mar/Apr 1971, pp. 246-252.
Ochsner, Alton: Lung cancer: the case against smoking, *The Nation,* May 23, 1953: pp. 431-432.
Ochsner, Alton: My first recognition of the relationship of smoking and lung cancer, *Preventive Medicine* 2:611-614, 1973.
Pearce, Janet: Cigarettes: is there a plot to keep you hooked? *Family Health* 10(6):20-23, 54, 1978.
Peto, Richard and Mathews, John: Smoking and cancer (letter). *Lancet* 2(7812):1126, May 19, 1973.
Schwartz, J.L. and Rider, G.: Review and evaluation of smoking control methods: the United States and Canada 1969-1977. USDHHS, Public Health Centers for Disease Control, Pub. #CDC 79-8369, 1978.
Soffer, Alfred, You've regressed a long way, baby, *Chest* 74(5):483-486, 1978.
Tobacco Institute: *About Tobacco Smoke,* 1979.
Tobacco Institute: *Before You Believe Half the Story, Get the Whole Story. Case in Point: The Public Smoking Issue,* Washington, D.C., 1981.
Tobacco Institute: *Fact or Fancy?,* 1979.
Tobacco Institute: *Federal Government Involvement in The Smoking and Health Controversy,* 1981.
Tobacco Institute: *Tobacco Industry Profile 1981,* 1981.
Werner, Carl Avery: The triumph of the cigarette, *American Mercury,* Dec. 1925, pp. 415-21.

Whiteside, Thomas: Selling death: cigarette ads in the magazines, *The New Republic* 164(13):15-17, 1971.

Wickstrom, Bo: *Cigarette Marketing in the Third World,* University of Gothenburg, Gothenburg, Sweden, 1979.

Wynder, Ernst L. and Cornfield, Jerome: Cancer of the lung in physicians, *The New England Journal of Medicine* 248:414-44, 1953.

Wynder, Ernst L.: Etiology of lung cancer, *Cancer* 30:1332-1338, 1972.

Wynder, Ernst L. and Graham, Evarts A.: Tobacco smoking as a possible etiologic factor in bronchiogenic carcinoma, *Journal of the American Medical Association* 143(4):329-338, 1950.

Index

"t" following a page number indicates a table concerning the subject.

A Dying Industry, description of, 128, 129
   tobacco industry and, 128-129
A Lifetime of Freedom from Smoking,
   American Lung Association, 215
Abbe, Robert, 55
Abbott, Twyman O., 48
Absenteeism, cigarette smoking and, 10
Acting prop, cigarette as, 75-76
Action for Smoking and Health, 115
Acupuncture, for smoking cessation, 219-220
Addiction, smoking as, 2
Additives, tobacco, 202-203
Adler, I., 55
Adolescents, cigarette advertising and, 106-107, 113
   smoking by, concern about, 120-121
   trends in, 143
Advertising, anti-smoking, 129-130
   by Agriculture Department in 1960s, 97-98
   by cigarette manufacturers, anything goes in, 183-184
   as economic clout, 5-6
   Cigarette Advertising and Labeling Act of 1971 and, 177
   countries banning, 188
   defense of, 5-6
   defensive tactics in, 92-94
   downplaying of health issue in, 184-185
   emphasis of, 177-181
   expenditures for, 177, 178t
   Federal Communications Commission and, 114
   Federal Trade Commission and, 75, 91-92, 94, 95, 103, 115-118
   in 1940s, 73-75
   in 1930s, 70
   in response to anti-smoking ads. 130-131
   in Third World countries, 169-170
   irresponsible, 111, 113
   purposes of, 181-182
   reassurance intended by, 182-183
   revenue from, 177, 179-180t
   self-regulation of, 106-107

   should be stopped, 144, 188-189
   suggestions for stopping, 189-190
   tar and nicotine and, 112
   to new markets, 185-186
   withdrawal of, 117
   by Tobacco Institute, 183-184
   in 1920s and 1930s, 57-59
   criticism of, 60
   of snuff, 135
   rail commuter, tobacco industry reactions to, 93-94
   reassurance of public through, 89
Advertising Age, True article and, 113
Agriculture Adjustment Act of 1933, 97
Alcohol, and smoking, related to cancer, 11
Amblyopia, nicotine, 54
America, smoking and, 207
American Cancer Society, anti-smoking commercials and, 115
   FreshStart by, 221
   Jack Anderson and, 20
   kit to guide quitting, 215
   prospective studies of smokers and nonsmokers, 85
   study commissioned by, 80
American Health Foundation, consultation on smoking cessation, 223
American Indians, smoking by, 29
American Journal of Cancer, linking cigarettes and lung cancer, 68
American Journal of Obstetrics and Gynecology, smoking and unborn child in, 86
   smoking during pregnancy and, 69
American Journal of Surgery, gravity of smoking in, 77
American Lung Association, *A Lifetime of Freedom from Smoking,* 215
   consultation on smoking cessation, 223
   costs of smoking to American businesses, 223
   Freedom From Smoking clinics by, 221-222
   Freedom from Smoking in 20 Days, 215
American Medical Association, education

234

leaflet by, 104
Federal Trade Commission hearings and, 104
American Mercury, in support of smoking, 63-64
smoking and health in, 79
American Pharmacy, study of filters in, 216
American Review of Tuberculosis, objection to smoking in, 68-69
American Tobacco Company, advertising by, 58
early history of, 42
Anatomy of Melancholy, by Robert Burton, 34
Anderson, Jack, 20
Anti-smoking advertising, 129-130
Anti-smoking campaign, by Joseph Califano, 121-123
Anti-smoking groups, after 1921, 51
in early 20th century, 49
Anti-tobacco movement, birth of, 36-37
Antismokers, Tobacco Institute and, 23-24
Arkin, Aaron, 69
Asbestos industry, and tobacco, 161-163
Asthma, exposure to cigarette smoke in, 194
Attitudes, about health and environment, 1
Auerbach, Oscar, 110,111
Austern, Thomas, 117

Bacall, Lauren, 64-76
Bans on cigarettes, in late 19th century, 46
Banzhaf, John F. III, 114, 115, 122, 159,163
Barron's, anti-smoking movement in, 209
Barron, Moses, 66-67
Behavior, automatic, cigarette smoking as, 3
Bekersky, Ihor, 220
Belli, Melvin, 155, 157
Benzo(a)pyrene, as cause of cancer, 81
Benzpyrene, in tobacco smoke, 193
Berger, Peter, 23-24
Berkson, Joseph, 77, 110
Bernays, Edward, 56-57
Beta-carotene, 20
Birth control pills, and cigarette smoking, 11
Black Lung Benefits Act of 1972, 160
Blasingame, F.J.L., 104
Blatnik, John A., 95
Blood vessel disease, cigarette-related, 10-11
Blum, Alan, 19
Bogart, Humphrey, 64, 76
Bogen, Emil, 68-69
Boling, James C., 31
Bonsack, James Albert, 42
Books, 230
Bouisson study, 38
Bourne, Peter, 121,122
Brand image, in advertising, 59-60
Brand names, early, 43
in 1940s, 72

Brill, A. A., 61-62
British Medical Journal, smoking and lung cancer in, 84-85
British Medical Research Council, study of smoking and cancer, 86
British-American Tobacco, memo, 183
British-American Tobacco (BAT), advertising by, 169
memo, 20, 183
Third World market and, 167
Bronchitis, chronic, smoking and, 12
Browder, Anne, 181, 186
Bruff, L. W., 181
Buchanan, Pat, 209
Buckley, William, 209
Burford, Thomas H., 155
Burney, Leroy, 95
Burton, Robert, 34
Business Week, cigarette sales prediction in, 132
tobacco industry reactions in, 89
Businesses, advice to, on smoking cessation, 223
Butler, C. S., 63

Cahan, William G., 110
Califano, Joseph, 20, 23, 121-123
California Medicine, tobacco propaganda in, 109-110
Camels, 52-53
advertising for, 58-59
Cameron, Charles S., 88
Campbell, Alexander, 69
Canadian Medical Association Journal, consolation for smokers in, 78
Cancer, alcohol and smoking related to, 11
and concept of "cause", 82
cigarette-induced, as dose-related, 17
well-known people dying of, 9
cigarette-related, 11
deaths attributable to, 12t
diet and, 20
environmental cause of, 81
esophageal, in cigar and pipe smokers, 140
laryngeal, 140
lung. *See* Lung(s), cancer
of mouth, smokeless tobacco and, 137-138
smoking as cause of, question concerning, 19
tobacco as cause of, 18th century report of, 35
Candy, -cigarette war, 62-63
cigarettes in place of, advertisements promoting, 62
Carbon monoxide, immediate effects of, 193
in tobacco smoke, 192-193
hazards of, 10
Carcinogens, in tobacco smoke, 193

Carmen, by Georges Bizet, cigarette in, 40, 41
Carter, Jimmy, 123-124
Cautions Against the Immoderate Use of Snuff, by John Hill, 35
Celebrezze, Anthony, 98
Celebrities, endorsement of cigarette smoking by, 57-58
Center for Science in the Public Interest, smoking and, 207-208
Charlotte Observer, The, facing up to facts in, 145
Chemicals, in cigarettes, 19
  in tobacco smoke, 192-193, 202
  in workplace, cancer and, 20
Chesterfields, 53
Chewing gum, nicotine, to aid smoking cessation, 217-218
Chewing tobacco, adverse effects of, 137
  advertising of, 135
  Indian studies on, 138
  juices of, 136
  reasons for popularity of, 137
  sales of, 136-137
  types of, 136
Chirurgical Observations, by P. Pott, cause of cancer in, 81
Cigar smokers, mortality rates for, 139
  sites of disease in, 139-140
Cigarette Advertising and Labeling Act of 1971, 177
Cigarette cards, 41
Cigarette companies. See Tobacco industry
Cigarette Country, 28-29
  tobacco advertising and, 70
Cigarette factories, growth of, 36
Cigarette holder(s), 66
Cigarette Labeling and Advertising Act, 107-108
Cigarette machine, early, 42
Cigarette manufacturers, major U.S., 225
Cigarette sales, in Third World countries, 167, 169
  Surgeon General's report and, 102
Cigarette smoke. See Smoke, tobacco
Cigarette(s), advantages of, 39
  advertising of. See Advertising, by cigarette manufacturers
  as "safer", as advertising message, 90
  as backfire of human innovation, 2
  as fire hazard, 151
  as hazardous product, as unique dilemma, 83-84
  bans on, in late 19th century, 46
  conspiracy of silence concerning, 141
  consumption, in 1950s, 94
  in Third World countries, 167
  increasing, 57
  trends in, 142-143, 142t
  demand for, in Europe in 1940s, 73
  facing facts concerning, 141
  full-flavored, in Third World countries, 170
  hand rolling of, 42
  image problems related to, 41
  ingredients of, tobacco industry and, 19
  king size, 60, 75
  low-tar, 124
  chemicals in, 203
  low-tar, low-nicotine. See Low-tar, low-nicotine cigarettes
  manufacture of, early, 40
  in United States, in late 19th century, 41, 42
  marketing, in Third World countries, 169
  leaders in, 56-57
  moods and messages conveyed by, 75-76
  opposition to, in early 20th century, 48-49
  in late 19th century, 41, 44-45
  production, decline of, in late 19th century, 46
  in early 20th century, 51-52
  in 1940s, 72-73
  promotion of, criticism of, 60
  in late 1920s and 1930s, 57-60
  in early 20th century, 53-54
  regulation of, state and local governments in, 108
  safety tax on, 165
  sales, in early 20th century, 49-50
  self-extinguishing, 164
  tobaccoless, 93
Cigarettes Advertising Code, 188-189
Civil Aeronautics Board, and smokers and nonsmokers, 191-192
Clements, Earl C., 105-106
Clinical Heart Disease, by S. A. Levine, smoking and health in, 78
Cocke, John Hartwell, 37
Cocoa, as additive, 203
Cognitive dissonance, 3
Cohen, Wilbur J., 112
Collins, Leroy, 106
Columbia Journalism Review, cigarette advertising in magazines in, 132
Columbus, Christopher, 29
Commonweal, criticism of cigarette advertising by, 60
Conflicts of interest, directors of cigarette companies and, 5
Congress, 97th, cigarette legislation considered in, 6
  cigarette policies of, 6-7
  actions concerning cigarettes, 95-96
  politics of tobacco in, complexity of, 7
  tobacco industry and, 103-104
  breaking grip of, 144
  vote-swapping in, 7
  warning labels and, 203-204
Congressmen, manipulation of, 6
Conservative leaders, smoking and, 209-210
Consumer advocacy movement, smoking and, 207-208
Consumer Product Safety Commission

Index 237

(CPSC), 164
Consumer Reports, cigarette health issue in, conservative attitude toward, 88-89
Consumers Union, smoking and longevity in, 70
Cooper vs. R. J. Reynolds Tobacco Company, 157
Coronary heart disease, exposure to cigarette smoke in, 193-194
Coronet, smoking message in, 87
Cort, Davis, 87
Costs, medical, of smoking-related diseases, 13
Coumarin, as cigarette additive, 203
Counterblaste to Tobacco, by King James I, 30-31
Creech, John L., 81-82
Cullman, Joseph, 130-131
Current History, objections to smoking in, 67
Current Opinion, objections to smoking in, 67

Darr, E. A., 89
Davidson, Ralph P., 131
Death in the West, airings in the U.S., 128
  blocking of showing of, 127
  reviews of, 127
Death(s), cancer, smoking and, 12t
  cigarette-related, cancers in, 12t
  incidence of, 10
  types of, 9-10
  well-known people dying of, 9
  concept of "excess", tobacco industry and, 24-26
  premature, in tobacco use, annual figures for, 25t
  diseases categories of 25t
DeBakey, Michael, 69
DeMorbis Artificum Diatriba, by Bernardino Ramazzini, 34-35
Denial, by cigarette pushers, 15-17
  second hand, 3
Dependence, cigarette smoking and, 2
  of suppliers of cigarette manufacturers, 4-5
Diet, and cancer, 20
Dillow, Gordon, 31, 52
Disability, bed, and cigarette smoking, 10
Disease, *See also Specific diseases*
  and smoking, R. J. Reynolds Tobacco Company and, 18
  environmental causes of, 81-82
  questions studies in, 82-83
  environmental factor and, causation between, criteria for, 17
  reduced incidence of, quitting and, 13-14, 17
Distraction, by cigarette pushers, 15-17
Dixon, W. E., 67
Dogs, smoking, studies of, 110-111
Doll, Richard, 84-85, 85
Dorn, Harold, 87

Doughboy, in cigarette ads, 57
Drath, Connie, 126
Drew, Elizabeth Brenner, 107-108
Drug dependence, cigarette smoking as, 2
Duke, James Buchanan, 42
Duke, Washington, 42, 43
Dupont, Robert, 202
Dwyer, William, 31, 126

Eastern Airlines, nonsmoking section on, 191
Education, cigarette smoking related to, 143
Emphysema, smoking and, 12
England, acceptance of cigarettes in, 40-41
  in early 17th century, export of tobacco to, 32
  introduction of cigarettes into, 40
English, John, 77
Environment, attitudes about, 1
Environmental factor, and disease, causation between, criteria for, 17
Epidemiology, and cause of disease, 16-17
  in study of cancer, 81-82
  standard questions in, 82-83
Esophagus, cancer of, in cigar and pipe smokers, 140
"Estimates Paper", by Joseph Califano, 20
Eve cigarettes, promotion of, 186
"Excess", concept of, tobacco industry and, 23
"Excess deaths", concept of, tobacco industry and, 24-26

Fact, introduction of, 185
Fallows, James, 131
Farmers, tobacco, 148
  in Third World countries, 168
Farming, tobacco, Third World countries and, 172
Federal Communications Commission, and air time for response to cigarette commercials, 114
Federal Trade Commission, advertising claims and, 75, 94, 95
  monitoring of, 91-92
  hearings of 1964, 104
  prohibition of cigarette advertising by, 115-118
  ruling of 1964, 104-105
  Surgeon General's report and, 103
  warning labels and, 203-204
  warning statement of 1967, 112
Feiser, Louis F., 101
Feldt, Robert, 79
Fillmore, Charles, 54
Filter(s), advantages claimed for, 94
  companies introducing, 94
  confusion of smokers concerning, 94-95
  for quitting of smoking, 216
  new Hi Fi, 90
Finkelstein, Max, 119

Fires, cigarette-related, 151
  incidence of, 13
  legal actions and, 163-164
Fisher, Irving, 65
Five-Day Plan, The, for quitting smoking, 220-221
Florio, James J., 7
Flue-curing, in Third World countries, 172
Food and Drug Administration, in 1960s, 98
  nicotine chewing gum and, 218
  smoking deterrents and, 216-217
Foote, Emerson, 57, 112, 181, 182
Fortas, Abe, 105
Fortune, survey on smoking, 70-71
Fountain, L. H., 114
France, cigarette in, 40
Frank, Stanley, 113
Freedom, choice of smoking as, 27
Freedom From Smoking clinics, by American Lung Association, 221-222
Freedom from Smoking in 20 Days, American Lung Association, 215
Freeman, Orville, 97
FreshStart, by American Cancer Society, 221
Freud, Sigmund, 47
Freidell, H. L., 77
Friedman, Milton, 209
Fritschler, A. Lee, 106
Froeb, Herman, 196

Galbraith, John Kenneth, 87
Gardner, Vincent, 220-221
Garfinkel, Lawrence, 195
Garner, Donald, 159, 160-161, 164
Gaston, Lucy Page, 45-46, 49, 50-51
Geography, as factor in tobacco issue, 6
Gitlitz, George, 152
Glyn, Elinor, 49
Gori, Gio Batta, 124
Government, attitudes toward cigarettes, in 1960s, 97-103
  recommendations for action of, toward Third World countries, 174, 175
  regulation, cigarette companies and, 5
Grace, Edwin, 77
Graham, Evarts A., 80, 84, 85, 96
Grant, Frederick Dent, 55
Grant, Ulysses S., 39
Greeley, Horace, 37
Green vs. American Tobacco, 156
Green, Samuel, 37
Greene, Bob, 201-202
Group methods, for smoking cessation, 220-222
Gumpert, Martin, 78

Haddon, Richard, 171
Hahn, Paul M., 89
Hammond, E. Cuyler, 85, 110
Hands, in cigarette ads, 58

Harding, Warren G., 51
Hawthorn Press, tobacco propaganda published by, 110
Hazard(s), cigarette smoking as, knowledge of, and methods of coping with, 3
  of cigarette smoking, avoidance of mention of, 3
Health, and tobacco, during 17th century, 33-34
  attitudes about, 1
  tobacco and, advertising promoting, 58
  as unique dilemma, 83-84
  cigarette advertising and, 184-185
  during 18th century, 34-35
  evidence against, 1920 to 1940, 66-71
  in 1960s, 97-108
  in early 20th century, 54
  in Third World countries, 171-172
  physicians' comments on, 78-79
  prospective studies in, 85
  retrospective studies in, 84-85
  studies of, in 1940s, 76-80
  in 1950s, 85-86
Health care, costs of, smoking and, 146-147, 147t
Health claims, in cigarette advertising, 58-59
Health warnings, on cigarette packages, 103, 107
  in Third World countries, 170
Heart disease, cigarette-related, 10-11
Heller, John R., 95
Helms, Jesse, 122, 147
Hettich, Arthur M., 131
Hill, A. Bradford, 84-85, 85
Hill, George Washington, 53, 60
  advertising by, 56
Hill, John, 35
Hirayama, Takeshi, 194-195
Hoffman, Frederick L., 67-68
Holder(s), cigarette, 66
Horn, Daniel, 85
Horrigan, Edward A., Jr., 186
Hubbard, Elbert, 46, 49
Hubbell, Charles, 45
Hudson vs. R. J. Reynolds Tobacco Company, 157
Humectant chemicals, in cigarettes, 203
Hunt, Jim, 23
Hyde, Leroy, 109-110
Hygeia, article promoting smoking in, 63
Hypnosis, as aid to quit smoking, 218-219
Hypnotist, choosing of, 219

Indians, American, smoking by, 29
Industry, tobacco, *See* Tobacco industry
Infant health, smoking and, 12-13
Infants, of smoking mothers, 196
Insurance benefits, to smokers and nonsmokers, 150
Insurance companies, discounts to nonsmokers by, 151

Index 239

Jacobson, Bobbie, 207
Jamestown, Virginia, export of tobacco from, 32
Japanese study, Tobacco Institute and, 26-27
Jenner, William, 82
Jesuits, smoking and, 93
Johnson, Lyndon, 102, 105
Johnson, Maurice N., 81-82
Johnson, Wingate, 63-64
Jones, Walter, 114
Jordan, David Starr, 49
Journal of the American Dental Association, smoking and cancer of oral cavity in, 77
Journal of the American Medical Association, advertisement for cigarettes in, 69
   consolation for smokers in, 78-79
   criticism of cigarette advertising by, 60
   results of first research on smoking in, 84
   smoking and lung cancer in, 69
   tobacco and health in, 77
Journal of the Medical Society of New Jersey, smoking by adolescents and, 69
Joyfullnewes oute of the newfounde worlds, 30

Kellogg, John Harvey, 67
Kennedy, John F., 99
Kent, advertising for, 111
   "micronite" filter of, 94
Kilpatrick, James, 208-209
King James I, 30-31
King size cigarettes, 60, 75
Kingsley, Charles, 34
Kools, marketing of, 185
Kornegay, Horace, 18, 105, 106, 185
Kraybill, Herman, 100
Kress, D. H., 49

Labels, warning. *See* Warning labels, on cigarette packages, 103, 107, 112
Ladies' Home Journal, cigarettes and disease in, 88
Lancet, message conveyed by smokers in, 182
   objections to smoking in, 67
   series of articles on tobacco in, in 1856 and 1857, 38
Lartigue vs. Liggett & Myers, 155
Larynx, cancer of, in cigar and pipe smokers, 140
Lasker, Albert, D., 56, 57
Lawsuits. *See* Legal action(s)
Legal action(s), Albright vs. R. J. Reynolds Tobacco Company, 157
   by widow of Glenn Crocker, 159
   cigarette-related fires, 163-164
   Cooper vs. R. J. Reynolds Tobacco Company, 157
   defense of "ignorance" in, 154
   Green vs. American Tobacco, 156

Hudson vs. R. J. Reynolds Tobacco Company, 157
   in asbestos-related illnesses, 161-162
Lartigue vs. Liggett & Myers, 155
   on behalf of family of John C. Galbraith, 157
Pritchard vs. Liggett & Myers, 156-157
Ross vs. Philip Morris and Co., 157
Legislation, needed, 8
   tobacco, in 1980s, 6-7
Lent, Richard, 191
Levin, Morton, 24-25
Levine, Samuel A., 78
Lickint, D., 68
Life expectancy, smoking and, 10
Little, Clarence Cook, 79, 92, 104
Lobbies, tobacco, confronting of, 8
Lobbyists, for tobacco industry, 105-106
Lobeline sulphate, as smoking deterrent, 216-217
Loomis, Russel, 68-69
Low-tar cigarettes, 124
   chemicals in, 203
Low-tar, low-nicotine cigarettes, advertising of, 130
   facts on, 133-134
   safety of, questioning of public on, 133
Lucky Strikes, 53
   advertising for, 59, 60, 62, 113
   George Washington Hill and, 56
   popularity of, five reasons for, 73-74
   promotion of, 60, 62
   in 1940s, 73-74
Lung(s), cancer, and smoking, claims of tobacco industry and, 19
   and smoking, in 1920 to 1940, 66-67, 69-70
   early 20th century perceptions of, 55
   epidemiological pattern of, 17
   Hirayama's study and, 194-195
   in asbestos exposure, 161-162
   mortality associated with, 11
   rate of, and improved diagnosis, 16
   smoking as cause of, 11
   diseases, smoking and, 12
   functional impairment of, in smokers and nonsmokers, 196
Lyatuu, Raymos, 171

Mack, Connie, 49
Magazines, cigarette advertising in, editors' comments on, 131-132
Mano, D. Keith, 136
Marketing, leaders in, 56-57
Marlboro, advertising for, 61
Marrin, Albert, 47
Matches, early, as dangerous, 44
   portable, introduction of, 43-44
Mattison, Henry W., 87
Mayan period, smoking in, 29
McDonald, John B., 77-78
McEvoy, J. P., 80

McNally, William, 68
Media support, of smoking, 63-66, 205
Mental dependence, 2
Mental deterioration, attributed to smoking, 49
Menthol, as cigarette additive, 203
Merrell Dow, nicotine chewing gum and, 218
Moffitt, Toby, 7
Mondale, Walter, 125
Moods, cigarette conveying, 75-76
MORE cigarettes, promotion of, 186
Morrison, Robert S., 120-121
Moss, Frank E., 108
Mother Jones, review of *Death in the West* in, 127
Mouth, cancer of, smokeless tobacco and, 137-138
Ms. Magazine, smoking and health in, 206
Muller, F. H., 70
Multiple Risk Factor Intervention Trial (MRFIT), Tobacco Institute and, 21
Murad IV of Turkey, 33
Music, popular, smoking as inspiration for, 47-48

Nader, Ralph, 207-208
Nation, cigarette health issue in, 88
 criticism of newsweeklies in, 87
 report of new filter in, 90
National Anti-Cigarette League, 51
National Cancer Institute, pamphlet for doctors, 218
 quit kits of, 215-216
National Council of Churches, smoking and, 207
National Organization for Women, smoking and, 206
Natural and Artificial Directions for Health, by Sir William Vaughn, 33
Neilson, William Allan, 71, 65
Neuberger, Maurine, 98, 99, 107
Neuberger, Richard L., 96
New England Journal of Medicine, and House hearings on cigarette advertising, 116-117
 nonsmokers exposed to smoke in, 120
New Republic, hostility toward cigarette industry in, 88
 in support of smoking, 63
 on smoking manners of women, 65-66
New York State Journal of Medicine, commercials featuring celebrities in, 135
New York Times, The, "balancing" of news article by, 87
 article on deceptive advertising in, 134
 claim for tobacco, 57
 in support of smoking, 63
 opposition to cigarettes in, 45
 promotion of cigarettes by, 54
 testimonials in, 64
Newhart, Bob, 130

Newsweek, causes of cancer in, 132
 cigarettes in injury in, 77-78
Nicot, Jean, 30
Nicotine, cravings, following cessation of smoking, 220
 dependence and, 2
 hazards of, 10
 in tobacco smoke, 192-193
Nicotine chewing gum, to aid smoking cessation, 217-218
Nicotinic amblyopia, 54
Nitrosamines, in tobacco smoke, 193
Nitrosonornicotine (NNN), 137
No-to-bac, 49
Non-Smokers Protection League of America, 49
Nonsmokers, activism of, tobacco industry and, 197
 and smoking cessation, 3-4
 direct costs of smoking paid by, 150-151
 insurance discounts to, 151
 rights of, 120, 191
 protection of, 144
 recommendations for protection of, 198-200
Nonsmokers' rights movement, of, 1910, 48
Nonsmoking section on Eastern Airlines, 191
Norr, Roy, 88
Norsigian, Judy, 206

O'Brasky, David M., 131
O'Neill, Tip, 125
Occupation, cigarette consumption related to, 143
Ochsner, Alton, 55, 69-70, 85, 155
Old Gold, *Reader's Digest* rating of, 74
Organizations, helpful, 228-229
Oski, Frank A., 223-224
Our Bodies Ourselves, smoking in, 206
Outlook, rights of non-smoker in, 48
Owen, Thomas, 203

Packages, cigarette, warning labels on, 103, 107, 112
Packaging, early, 43
 of Duke cigarettes, 43
 of Lucky Strikes, 73
Paget, Sir James, 38
Passive smoking, 192
 effects of, 194-196
Pearl, Raymond, 70
Pearson, Drew, 108
Pease, Charles G., 49
Peptic ulcer disease, 12
Pershing, General John J., 50
Pertschuk, Michael, 117, 204-205
Philip Morris, advertising claims by, 75
 advertising for, 58, 59, 90
 college campus promotion of, 90
 Death in the West and, 127-128
 directors of, 5

Index 241

Physical dependence, 2
Physical deterioration, attributed to smoking, 49
Physicians, opposed to smoking, in early 20th century, 48
Pinney, John M., 163
Pipe smokers, mortality rates for, 139
  sites of disease in, 139-140
Plague of 1665, smoking during, 33-34
Polin, William, 2
Political coalition, tobacco, 106
Politics, cigarette, in 1980s, 6-8
Pope Pius XII, 93
Pott, Percival, 81
Pregnancy, smoking during, 12, 196
Premium coupons, 75
President's Advisory Committee for Women, smoking and, 206-207
Press, cigarettes and, 86-89
Primary Malignant Growths of the Lung and Bronchi, by I. Adler, 55
Prince Albert tobacco, 52
Printer's Ink, survey of ad men and smoking, 111
Pritchard vs. Liggett & Myers, 156-157
Productivity, and smoking, 49
Promotion, by cigarette manufacturers, as economic clout, 5-6
  of social acceptance of cigarettes, 6
  of cigarettes, criticism of, 60
  in 1920s and 1930s, 57-60
Protests, lack of, factors causing, 1-2
Psychological blackout, 3
Psychological mechanisms, and smoking cessation, 3-4
Psychological reactions in cigarette smoke exposure, 194
Public Health Service, actions by, 111-112
  report of 1957, 98
Publications, 230-233
  self-help, for quitting smoking, 215-216
Pullen, David, 162

Quinoline, in tobacco smoke, 193
Quitting. *See* Smoking, quitting of

R. J. Reynolds Tobacco Company, 4, 5, 18
Raleigh, Sir Walter, 30
Ramazzini, Bernardino, 34-35
Reader's Digest, ads for articles in, tobacco industry reactions to, 93-94
  cigarettes and lung cancer in, 88
  crusade against tobacco in, 79-80
  educational campaign by, 65
  middle-of-the-road position of, 87
  on smoking manners of women, 66
  research laboratory findings by, 74
Reagan administration, tobacco industry influence on, 7-8
Reagan, Ronald, 208
Reddan, William G., 93
Religious institutions, smoking and, 207

Research, by tobacco companies, 23
Reynolds, Richard Joshua, 52
Richmond, Julius, 124-125, 133-134, 202-203
Riis, Roger William, 87
Rimer, Irving, 127
Robinson, Jackie, 112
Roffo, A. H., 68
Rogers, Tom, 33-34
Rolfe, John, 32
Roosevelt, Mrs. Franklin D., 61
Roper Organization, report for Tobacco Institute, 186-188
Rosenthal, L. M., 77
Ross vs. Philip Morris and Co., 157
Royal College of Physicians in London, "Smoking and Health" report of, 100
Rush, Benjamin, 35

Safire, William, 209
Salem, advertising-marketing plan for, 182-183
San Francisco Women's Health Collective, smoking and, 206
Sanford, Terry, 23
Sanguinetti, H. H., 68
Schick Clinics, 222
Schneider, John, 87
Schrumpf-Pierron, Pierre, 67
Schumann-Heink, Madame, 57
Science, cigarette sales and Surgeon General's report in, 102-103
  smoking and longevity in, 70
Science Digest, smoking and health in, 78, 79
Science News Letter, smoking and cancer in, 69
Scientific study group, requested by U.S. Surgeon General, conclusions of, 86
Scientists, tobacco industry and, 21
Self-help publications, for quitting smoking, 215-216
Seltzer, Carl C., 21
Senators, manipulation of, 6
Senevitatne, Gamini, 169
Seventh-day Adventists, smoking and, 207
  study of cancer and heart disease in, 86
Sexual themes, in music, smoking and, 47-48
Sharlit, Herman, 69
Sherman Anti-Trust Act, cigarette companies and, 72
Shew, Joel, 37-38
Short, James, 70
Shortway, Richard A., 131
Signs, "No Smoking", 198, 199
Silver acetate, as smoking deterrent, 217
Simpson, Winea J., 86
Skoal Bandits, 135
Small, Edward Featherston, 43
Smith, James F., 137-138
Smith, Martin, 127

Smith, R. D., 132
Smoke, cigarette, components of, dangers of, 10
  inhalation of, 28
  tobacco, contents of, 192-193
  immediate effects of, 193-194
  recommendations for action against, 198-200
  sensitivity to, 193
Smoke Watchers, 222
SmokEnders, 222
Smokers, "creed" for, 64-65
  below "age of consent", 159-160
  characteristics of, and disease, 16
  facts for, need for, 144
  reactions of, to Surgeon General's report, 102-103
  readjustments of burden to, 151
  recommendations for action by, 210-211
  should pay own way, 143-144
Smoking, answers to "why?", 201-202
  as addiction, 202
  as drug dependence, 2
  as everybody's business, 192-193
  asbestos-related disease and, 161-162
  by parents, children in, 196-197
  cessation method, choosing of, 213
  during pregnancy, 12, 196
  economic burden of, recommendations for shifting, 152-153
  efforts to deter, in early 17th century, 31, 33
  habit, creation of, 169
  health care costs and, 146-147, 147t
  history of, 28-29
  in early 17th century, social levels of, 31-32
  in public places, 192
  recommendations for actions against, 199-200
  tobacco industry and, 197
  in Third World countries, health consequences of, 171, 172
  in workplace, 192
  long-term effects of, 194-196
  on commercial airlines, 191-192
  passive, 192
  effects of, 194-196
  plan to discourage, 152
  quitting of, acupuncture for, 219-220
  and reduced incidence of disease, 13-14, 17
  group methods for, 220-222
  hypnosis for, 218-219
  medical supervision for, 218
  nicotine chewing gum for, 217-218
  over-the-counter products for, 216-217
  reasons for, 212
  risk of cancer associated with, 13-14
  self-help publications for, 215-216
  success rates in, 222-223
  successful, 212-213
  weight gain following, 214-215
  research on, prospective studies in, 85
  retrospective studies in, 84-85
  risks of, 9-14
  increasing public awareness of, 204-205
  lack of awareness of, 201
  minimal news coverage of, reasons for, 205
  overall, 10
  quantifying of, 13-14, 13t
  studies in dogs, 110-111
  withdrawal symptoms, 213-214
Smoking cessation clinics, 152
Smoking deterrents, 216-217
Snow, John, 81
Snuff, adverse effects of, 137
  advertising of, 135
  reasons for popularity of, 137
  sales of, 136-137
  types of, 136
  use by young males, 135
  victims of, in 18th century, 35
Sobel, Robert, 44
Social acceptance of cigarettes, promotion of, 6
Songs, inspired by cigarettes, 47
Spain, cigarette in, 40
Spitters, public, legislation against, 197-198
St. John, Jeffrey, 209
Standard Asbestos Manufacturing and Insulating Company, 162
Statistical relationships, and cause of disease, 16-17
Sterling, Theodor D., 21
Stewart, W. Blair, 69
Stopping smoking. *See* Smoking, quitting of
Studies, of harmful effects of tobacco, in early 20th century, 54-55
  tobacco industry criticisms of, 26-27
Sudden Infant Death Syndrome, 196
Suits. *See* Legal action(s)
Superman II, advertising in, 186
Suppliers, of cigarette manufacturers, dependency of, 4-5
Surgeon General's committee, 99
  meetings of, 100
  members of, 99-100
Surgeon General's report, actions of Luther Terry following, 101
  conclusions of, 101
  Federal Trade Commission and, 103
  quote from, used by advocates of smoking, 16-17
  reactions of Lyndon Johnson to, 102

Talman, William, 115
Tar, in tobacco smoke, 192-193
Tareyton, advertising for, 59
Task Force on Smoking and Health, 112
Tax, "health", on cigarettes, 151
  excise, 6

Index   243

on cigarettes, dependency of government on, 52
safety, on cigarettes, 165
Taylor, Peter, 127,128
Ted Bates and Company, Inc., report of, 184-186
Teenagers. See Adolescents
Television, banning of cigarette advertising on, 117-118
Terry, Luther, 99, 100, 101, 103
Test protocol, in test of cigarettes' harmfulness, 17-18
Testimonials, on smoking, 64
They Satisfy, by Robert Sobel, 44
Third World Countries, cigarette advertising in, 169-170
 cigarette consumption in, 167
 full-flavored cigarettes in, 170
 lure of revenues from tobacco products in, 168
 recommendations for action in, 174-176
 smoking in, health consequences of, 171, 172
 tobacco industry and, 166
 tobacco production in, dream of export in, 173
 hidden costs of, 173
 new technologies for, 174
Thoughts and Stories for American Lads, by George Trask, 36
Throat, irritation of, cigarette advertising and, 58
Tobacco, addictive nature of, 158-159
 and health, during 17th century, 33-34
 during 18th century, 34-35
 in early 20th century, 54
 as cause of cancer, 18th century report of, 35
 as colonial export product, 32
 as essential crop, 72
 as "golden token", 34
 as growth industry, in late 19th century, 39
 as New World plant, 29
 benefits of, during World War I, 50
 chewing. See Chewing tobacco
 dangers of, studies of, in early 20th century, 54-55
 dependency, as "habituation", 158
 health and. See Health, tobacco and
 in Food for Peace Program, 167-168
 in United States in 19th century, 35-39
 inhalation of, 28
 New York Times claim for, 57
 opposition to, beginnings of, 36-37
 price support of, in 1960s, 97
 price support program, 147-148
 production, in developed world, 173
 in Third World countries, dream of export in, 173
 hidden costs of, 173
 new technologies for, 174

smokeless. See Chewing tobacco; Snuff;
 "Smokeless" tobacco use of, as social activity, 36
 forms of, 28
 history of, 29
Tobacco Action Network (TAN), 15
Tobacco farming, Third World countries and, 172
Tobacco industry, "official dishonesty" of, 92
 A Dying Industry and, 128-129
 and "freedom of choice", 201-202
 and activism by nonsmokers, 197
 and Congress, breaking of grip of, 144
 and ingredients of cigarettes, 19
 as big business, 4
 as heavily taxed, 26
 as old as America itself, 26
 as punching bag, 19
 asbestos industry and, 161-163
 attitude of, toward smoking and health claims, 19-20
 beneficiaries in, 149
 campaign to discredit opponents, 125-126
 concept of "excess", 23
 contribution of, to economy, 148-149
 to Gross National Product, 149
 corporate directors and, 5
 diversification in, 118-119
 economic clout of, advertising and promotion and, 5-6
 in 1940s, 72
 in 1960s, 97
 influence of, on Reagan administration, 7-8
 lobbying groups of, 109
 lobbyists for, 105-106
 mandates for action against, 143-145
 methods of, for dealing with anxiety concerning cigarettes, 20
 network of, 225-227
 political clout of, family of companies and, 4
 manipulation of legislators and, 6
 political allies and, 5
 power over suppliers and, 4-5
 protection of self by, 126-129
 psychological blackout and, 3
 reactions of, to Federal Trade Commission hearings, 105-106
 to medical data on cigarettes, 89
 to Surgeon General's report, 102
 readjustment of burden to, 151
 reassurance of public by, 89-91
 research by, 23
 responses of, to anti-smoking advertising, 130-131
 self-regulation of, 106-107
 should have responsibility for damage, 144
 strength of, 2

studies sponsored by, 110
subsidization of, 147-148
thoroughness of, 15
U.S. Congress and, 103-104
Tobacco Industry Research Committee, 92-93
Tobacco Institute, ads directed at journalists, 183-184
and antismokers, 23-24
and cancerphobia, 21-22
and rates of three kinds of cancer, graph of, 21, 22
flag-wrapping by, 91
in denial of health hazard from smoking, 126
in support of smoking, 14
inquiry to, on product safety criteria, 18
"Japanese study" and, 26-27
misinterpretation of research data by, 21
on purposes of cigarette advertising, 181
propaganda of, 15
propaganda tactics of, 109-110
reaction to tobaccoless cigarette, 93
reactions of, to Surgeon General's report, 102
Roper Organization report for, 186-188
Tobacco Leaf, tobaccoless cigarette and, 93
Tobacco Observer, 15
articles in, 20
headlines in, 20-21
purposes of cigarette advertising in, 181
tobacco industry and nonsmokers' rights in, 197
Tobacco spitting contest, 135-136
Tobacco: Its History, Nature and Effects on the Body and Mind, by Joel Shew, 37-38
Tobacco: The Bane of Virginia Husbandry, by John Hartwell Cocke, 37
Tobey, James, 60
Trask, Reverend George, 36
True, smoking article planted in, 113
Tunney, Gene, 80
Twain, Mark, 141
Tylecote, Frank E., 67

U.S. Congress. *See* Congress
U.S. government. *See* Government
U.S. News and World Report, cigarette health issue in, 88
Ulcers, peptic, 12
Ulfelder, Howard, 81
United Nation's Food and Agriculture Organization, recommendations for action by, 174

United States Tobacco Journal, Jesuits and tobacco in, 93

Vanguard, 93
Vaughn, Sir William, 33
Virginia Company, and smoking as fad, 32

Wagner, David, 69
Wagner, Susan, 28-29, 70
War, cigarettes in time of, 72
Warning labels, brainwashing and, 7
cigarette legislation and, 6-7
on cigarette packages, 103, 107, 112, 203-204
Federal Trade Commission recommendations for, 203-204
liability of manufacturers and, 158
Washington Monthly, advertising space to cigarette manufacturers in, 131
Weight gain, following cessation of smoking, 214-215
Weiss, David, 137
Werner, Carl Avery, 64-65
White, James, 196
Wicker, Tom, 213
Wieckert, Steven, 148
Willius, Frederick, 77
Withdrawal symptoms, 213-214
Wolfe, Sidney, 208
Women, cigarette advertising aimed at, 186
cigarette advertising appealing to, 61
cigarette advertising featuring, 61, 62
cigarette smoking by, 11
in early 20th century, 51
in promotion of cigarettes, 62
smoking by, and independence of, 61
in 19th century, 36
smoking manners of, 65-66
World Bank, promotion of tobacco by, 168
World Health Organization (WHO), recommendations of, on smoking and tobacco production, 175
World Tobacco, demand for cigarettes in Third World in, 169
full-flavored cigarettes in Third World countries in, 170
World War I, cigarette use during, 50
Writing, on smoking, at turn of century, 47-49
Wynder, Ernst L., 17, 80, 84, 85, 86, 155

Young, George, 181-182